DICTIONARY OF ENVIRONMENT

DICTIONARY OF ENVIRONMENT

K.R. Gupta
Ashish Kumar Gupta
Suriti Gupta

ATLANTIC
PUBLISHERS & DISTRIBUTORS (P) LTD

Published by

ATLANTIC

PUBLISHERS & DISTRIBUTORS (P) LTD

B-2, Vishal Enclave, Opp. Rajouri Garden,
New Delhi-110027
Phones : +91-11-25413460, 25429987, 25466842

Sales Office
7/22, Ansari Road, Darya Ganj,
New Delhi-110002
Phones : +91-11-23273880, 23275880, 23280451
Fax : +91-11-23285873
web : www.atlanticbooks.com
e-mail : info@atlanticbooks.com

Chennai Branch
5, Nallathambi Street, Wallajah Road,
Chennai-600002
Phones : +91-44-64611085, 32413319
e-mail : chennai@atlanticbooks.com

Printed in India
at Nice Printing Press, Delhi

Preface

Environment can be defined as the natural world in which people, animals and plants live. It includes the biotic components like flora and fauna, the abiotic components such as air, water, soil and minerals, and the climatic conditions that affect the behaviour and development of all living beings on the earth.

Man is both the creature and moulder of his environment, which gives him physical sustenance and the opportunely for intellectual, moral, social and spiritual growth. In the long and tortuous evolution of the human race on this planet, a stage has been reached when, through the rapid acceleration of science and technology, man has acquired the power to transform his environment in countless ways and on an unprecedented scale. Both aspects of man's environment, the natural and the man-made, are essential to his well-being.

Today, the protection and improvement of environment is a major issue that affects the life of human beings and the socio-economic development of every nation in the world. Organic pollutants like industrial chemicals, pesticides and by-products of combustion are linked with a wide range of impacts on human health, wildlife and vegetation. Release of low-level radioactive waste at the sea and oil spills are playing havoc with the marine environments. Realising the importance of environment and considering the need to curb the unsustainable practices that are polluting the environment, the United Nations in its Conference on the Human Environment, at Stockholm, in 1972, emphasised on a common outlook and common principles to inspire and guide the peoples of the world to protect the environment. The efforts are continuing into the new millennium as the problems relating to environment such as global warming, depletion of ozone layer and damaging of natural habitats have aggravated.

Any action plan for the protection of environment necessitates the knowledge of various terms relating to environment and their exact connotations. Even otherwise, environment is such a vast phenomenon that its different aspects cannot be fully understood until we have grasped its terminology. There is a world-wide unanimity in the opinion that a perceptible change in the preservation of environment cannot be brought unless there is an awareness among the people about the damaging effects of unsustainable human activities. This *Dictionary of Environment* ably serves both these purposes—of providing valuable information about environment through a comprehensive terminology—and creating an awareness at the grassroots level by explaining the implications of various human activities.

The book will be highly useful for the students and teachers of Earth Science, Environment, Geography and Geology. The aspiring candidates of various competitive examinations will find it immensely helpful. It will greatly benefit the researchers and scholars in these fields. Common readers and those concerned with environment will find it highly informative.

K.R. Gupta
Ashish Kumar Gupta
Suriti Gupta

A

Abatement. Diminishing the degree or intensity of, or eradicating, pollution.

Abatement Debris. Waste obtained from remediation activities.

Abiotic. Non-living. Compare *biotic*.

Ablation (glacial). All processes, which include melting, evaporation (sublimation), wind erosion, and calving (breaking off of ice masses), that remove snow or ice from a glacier or snowfield. The term also refers to the amount of snow or ice removed by these processes.

Abrupt Climate Change. A change in climate over a widespread area that takes place so rapidly and unexpectedly that human and natural systems have difficulty in adapting to it. An abrupt climate change occurs on the scale of decades, rather than centuries, and persists for years.

Abscission. Shedding by a plant of its parts, such as leaves, flowers, fruits, or seeds. The process is regulated by the plant hormone abscisic acid.

Absorbed Dose. In exposure assessment, the quantity of a substance that penetrates an exposed organism's absorption barriers (*e.g.* skin, lung tissue, gastrointestinal tract) through physical or biological processes.

Absorption. The uptake of water, other liquids, or dissolved chemicals by a cell or an organism (as tree roots absorb dissolved nutrients in soil).

Absorption Barrier. Any of the exchange sites of the body that permit uptake of various substances at different rates (*e.g.* skin, lung tissue, and gastrointestinal-tract wall).

Absorption Coefficient. A measure of the amount of radiant energy, incident normal to a planar surface, that is absorbed per unit distance or unit mass of a substance.

Absorption Field. A system of properly sized and constructed narrow trenches partially filled with a bed of washed gravel or crushed stone into which perforated or open joint pipe is placed. The discharge from the septic tank is distributed through these pipes into trenches and surrounding soil. While seepage pits normally require less land area to install, they should be used only where absorption fields are not suitable and well-water supplies are not endangered.

Absorption of Radiation. The uptake of radiation by a solid body, liquid or gas. The absorbed energy may be transferred or re-emitted.

Accident Site. The location of an unexpected occurrence, failure or loss, either at a plant or along a transportation route, resulting in a release of hazardous materials.

Acclimated Culture. Culture of microbes that has been developed to use a particular substrate or waste-water.

Acclimated Sludge. A culture-activated sludge from a municipal plant that is gradually fed increasing amounts of the substrate or waste-water until the microbes using these food substances have grown in abundance.

Acclimation (acclimatization). Change that occurs in an organism to allow it to tolerate a new environment.

Accretion. A phenomenon consisting of the increase in size of particles by the process of external addition; for example, water to salt particles.

Accumulation (glacial). All processes, which include snowfall, condensation, avalanching, snow transport by wind, and freezing of liquid water, that add snow or ice to a glacier, floating ice, or snow cover. The term also includes the amount of snow or other solid precipitation added to a glacier or snowfield by these processes.

Acid. A corrosive solution with a pH less than 7.

Acid Aerosol. Acidic liquid or solid particles small enough to become airborne. High concentrations can irritate the lungs and have been associated with respiratory diseases like asthma.

Acid Deposition. A complex chemical and atmospheric process whereby recombined emissions of sulfur and nitrogen compounds are redeposited on earth in wet or dry form.

Acid Mine Drainage. Drainage of water from areas that have been mined for coal or other mineral ores. The water has a low pH because of its contact with sulfur-bearing material and is harmful to aquatic organisms.

Acid Neutralizing Capacity. Measure of ability of a base (*e.g.* water or soil) to resist changes in pH.

Acid Rain. Rainwater that has an acidity content greater than the postulated natural pH of about 5.6. It is formed when sulfur dioxides and nitrogen oxides, as gases or fine particles in the atmosphere, combine with water vapor and precipitate as sulfuric acid or nitric acid in rain, snow, or fog. The dry forms are acidic gases or particulates.

Acid Soil. Soil with a pH value < 6.6.

Acid Solution. Any water solution that has more hydrogen ions (H^+) than hydroxide ions (OH^-); any water solution with a pH less than 7.

Acidic. See *acid solution.*

Acidity. The capacity of water or waste-water to neutralise bases. Acidity is expressed in milligrams per liter of equivalent calcium carbonate. Acidity is not the same as pH.

Acidity Profile. The acid concentration in ice core layers as a function of depth as determined from electrical measurements. The magnitudes of some volcanic eruptions in the Northern Hemisphere have been estimated from the acidity of annual layers in ice cores taken in Greenland. This methodology is sometimes referred to as acidity signal or acidity record.

Acoustics. The field of science that includes the study of sound generation and transmission.

Activated Carbon. A highly adsorbent form of carbon used to remove odors and toxic substances from liquid or gaseous emissions. In waste treatment, it is used to remove dissolved organic matter from waste drinking water. It is also used in motor vehicle evaporative control systems.

Activated Sludge. Product that results when primary effluent is mixed with bacteria-laden sludge and then agitated and aerated to promote biological treatment, speeding the breakdown of organic matter in raw sewage undergoing secondary waste treatment.

Activated Sludge Process. Biological waste-water treatment process in which a mixture of the waste-water and activated sludge is aerated in a reactor basin or aeration tank. Active biological solids bio-oxidise the waste matter and the biological solids are removed by secondary clarification or final settling.

Activation Energy. Amount of energy required to bring all molecules in one mole of a substance to their reactive state at a given temperature.

Activator. A chemical added to a pesticide to increase its activity.

Active Ingredient. In any pesticide product, the component that kills, or otherwise controls, target pests. Pesticides are regulated primarily on the basis of active ingredients.

Adaptation. The adjustment of an organism or population to a new or altered environment through genetic changes brought about by natural selection.

Adaptive Radiation. The process that occurs when a species enters a new habitat that has unoccupied niches and evolves into a group of new species, each adapted to one of these niches.

Adiabatic Process. A thermodynamic change of state of a system such that no heat or mass is transferred across the boundaries of the system. In an adiabatic process, expansion always results in cooling, and compression in warming.

Adiabatic Warming. See *adiabatic process.*

Adsorption. Removal of a pollutant from air or water by collecting the pollutant on the surface of a solid material; *e.g.*, an advanced method of treating waste in which activated carbon removes organic matter from waste-water.

Advanced Waste-water Treatment. Any treatment of sewage that goes beyond the secondary or biological water treatment stage and includes the removal of nutrients such as phosphorus and nitrogen and a high percentage of suspended solids.

Advection. The predominately horizontal large-scale movement of air that causes changes in temperature or other physical properties. In oceanography, advection is the horizontal or vertical flow of sea water as a current.

Aerated Lagoon. A holding and/or treatment pond that speeds up the natural process of biological decomposition of organic waste by stimulating the growth and activity of bacteria that degrade organic waste.

Aeration. A process which promotes biological degradation of organic matter in water. The process may be passive (as when waste is exposed to air), or active (as when a mixing or bubbling device introduces the air).

Aerobacter Aerogenes. One of the species of bacteria in the coliform group.

Aerobe. A microorganism that lives and grows in the presence of free gaseous oxygen (O_2).

Aerobic. A life or process that occurs in and is dependent upon oxygen.

Aerobic Bacteria. Bacteria requiring free molecular oxygen for their life processes.

Aerobic Digestion. Digestion of suspended organic matter by aerobic microbes.

Aerobic Treatment. Process by which microbes decompose complex organic compounds in the presence of oxygen and use the liberated energy for reproduction and growth.

(Such processes include extended aeration, trickling filtration, and rotating biological contactors.)

Aerosol. Particulate matter, solid or liquid, larger than a molecule but small enough to remain suspended in the atmosphere. Natural sources include salt particles from sea spray, dust and clay particles as a result of weathering of rocks, both of which are carried upward by the wind. Aerosols can also originate as a result of human activities and are often considered pollutants. Aerosols are important in the atmosphere as nuclei for the condensation of water droplets and ice crystals, as participants in various chemical cycles, and as absorbers and scatters of solar radiation, thereby influencing the radiation budget of the Earth's climate system.

Afforestations. Planting of new forests on lands that have not been recently forested.

Afterburner. A device, which includes an auxiliary fuel burner and a combustion chamber, in which combustible air contaminants are incinerated.

Agar. Complex polysaccharide derived from certain marine algae that is a gelling agent for solid or semisolid microbiological media. Agar consists of about 70 per cent agarose and 30 per cent agaropectin. Agar can be melted at temperature above 100°C; gelling temperature is 40–50°C.

Agent. Any physical, chemical, or biological entity that can be harmful to an organism (synonymous with stressors).

Agglomeration. In meteorology, the process by which *precipitation* particles grow larger by collision or contact with *cloud* particles or other precipitation particles.

Agricultural Pollution. Farming wastes, including runoff and leaching of pesticides and fertilizers; erosion and dust from plowing; improper disposal of animal manure and carcasses; crop residues, and debris.

Agricultural Waste. Poultry and livestock manure, and residual materials in liquid or solid form generated from the production and marketing of poultry, livestock or fur-bearing animals; also includes grain, vegetable, and fruit harvest residue.

Agroecosystem. Land used for crops, pasture, and livestock; the adjacent uncultivated land that supports other vegetation and wildlife; and the associated atmosphere, the underlying soils, groundwater, and drainage networks.

Agung. Active volcano 10,380 ft (3,141 m) high in Bali, Indonesia. Last eruption was in 1964.

Air Binding. Situation where air enters the filter media and harms both the filtration and backwash processes.

Air Carrier. An operator (*e.g.* airline) in the commercial system of air transportation consisting of aircraft that hold certificates of, Public Convenience and Necessity, issued by the Department of Transportation, to conduct scheduled or non-scheduled flights within the country or abroad.

Air Contaminant. Any particulate matter, gas, or combination thereof, other than water vapor.

Air Monitoring. The continuous sampling and measuring of the quantity of air pollutants present in the atmosphere.

Air Pollutant. Any substance in air that could, in high enough concentration, harm man, other animals, vegetation, or material. Pollutants may include almost any natural or artificial composition of airborne matter capable of being airborne. They may be in the form of solid particles, liquid droplets, gases, or in combination thereof. Generally, they fall into two main groups: (1) those emitted directly from identifiable sources and (2) those produced in the air by interaction between two or more primary pollutants, or by reaction with normal atmospheric constituents, with or without photoactivation. Exclusive of pollen, fog, and dust, which are of natural origin, about 100 contaminants have been identified. Air pollutants are often grouped in categories for ease in classification; some of the categories are: solids, sulfur compounds, volatile organic chemicals, particulate matter, nitrogen compounds, oxygen compounds, halogen compounds, radioactive compound, and odors.

Air Pollution. The presence of contaminants or pollutant substances in the air that interfere with human health or welfare, or produce other harmful environmental effects. Contaminants in the air we breathe come mainly from manufacturing industries, electric power plants, exhaust from automobiles, buses and trucks.

Air Pollution Episode. A period of abnormally high concentration of air pollutants, often due to low winds and temperature inversion, that can cause illness and death.

Air Pressure. The weight of the atmosphere over a particular point, also called barometric pressure. Average air exerts approximately 14.7 pounds (6.8 kg) of force on every square inch (or 101,325 newtons on every square meter) at sea level.

Air Quality Standards. Levels of air pollutants that delineate acceptable levels of pollution over a particular time period. Valuable because they are often tied to emission standards that attempt to control air pollution.

Air Stripping. A treatment system that removes volatile organic compounds (VOCs) from contaminated groundwater or surface water by forcing an airstream through the water and causing the compounds to evaporate.

Airborne Fraction. The portion of CO_2 released from all energy consumption and land use activities that remains in the atmosphere as opposed to the amounts absorbed by plants and oceans. How the world's total carbon is partitioned among the oceanic, terrestrial, and atmospheric pools is determined by complex biogeochemical and climatological interactions.

Airborne Particulates. Total suspended matter found in *the atmosphere* as solid pieces or liquid droplets. Airborne particulates include windblown dust, emissions from industrial processes, smoke from the burning of wood and coal, and the exhaust of motor vehicles.

Airborne Release. Release of any pollutant into the air.

Airborne Sound. The sound that reaches the point of measurement travelling through the air.

Airmass. A widespread body of in *the atmosphere* that gains certain meteorological or polluted characteristics while set in one location. The characteristics can change as it moves away.

Albedo. Refers to the ratio of light from the sun that is reflected by the Earth's surface to the light received by it. It is the fraction of the total solar radiation incident on a body that is reflected by it. Albedo can be expressed as either a percentage or a fraction of 1. Unreflected light is converted to infrared radiation (*i.e.*, heat), which causes atmospheric warming. Thus, surfaces with a high albedo generally contribute to cooling, whereas surfaces with a low albedo generally contribute to warming. Changes in land use that significantly alter the characteristics of land surfaces can therefore influence the climate through changes in albedo. Snow covered areas have a high albedo (up to about 0.9 or 90%) due to their white color, while vegetation has a low albedo (generally about 0.1 or 10%) due to the dark color and light absorbed for photosynthesis. Clouds have an intermediate albedo and are the most important contributor to the Earth's albedo. The Earth's aggregate albedo is approximately 0.3.

Alfisols. An order of soils with a medium-to-high base supply, horizons of clay accumulation, and gray-brown surface horizons.

Algae. Simple rootless plants that grow in sunlit waters in relative proportion to the amounts of *nutrients* available. They can affect water quality adversely by lowering the dissolved oxygen in the water. They are food for fish and small aquatic animals.

Algal Blooms. Sudden spurts of algal growth that can indicate potentially hazardous changes in local water chemistry.

Alkaline. The condition of water or soil which contains a sufficient amount of alkali substance to raise the pH above 7.0.

Alkalinity. A pressure- and temperature-independent property of seawater that determines in part the carbon content of seawater. Carbonate alkalinity is the sum of the concentration of bicarbonate plus two times the concentration of the carbonate ions. Total alkalinity is the amount of acid required to bring seawater to a pH at which all dissolved inorganic carbon becomes freely exchangeable. The alkalinity of the oceans is determined with potentiometric or normal titration techniques that detect and measure the presence of bicarbonate, carbonate, and borate ions.

Alliance of Small Island States (AOSIS). The group of Pacific and Caribbean nations who call for relatively fast action by developed nations to reduce greenhouse gas emissions. The AOSIS countries are particularly vulnerable to the potential adverse consequences of climate change and concerned with the effects of rising sea levels and increased storm activity predicted to accompany global warming.

Allocation. Under an emissions trading scheme, permits to emit can initially either be given away for free, usually under a 'grandfathering' approach based on past emissions in a base year or an 'updating' approach based on the more recent emissions. The alternative is to auction permits in an initial market offering.

Allochthonous Flora. Organisms that are not indigenous to the soil but that enter soil by precipitation, diseased tissues, manure and sewage. They may persist for some time but do not contribute in a significant way to ecologically significant transformations or interactions.

Alluvial. Relating to and/or sand deposited by flowing water.

Alternative Energy. Energy derived from nontraditional sources (*e.g.* compressed natural gas, solar, hydroelectric, wind).

Altithermal Period. A period of high temperature, particularly the one from 8000 to 4000 B.P. (before the present era), which was apparently warmer in summers, as compared with the present, and with the precipitation zones shifted poleward. Also called the *hypsithermal period.*

Ambient Air. Any unconfined portion of the atmosphere: open air, surrounding air.

Ambient Measurement. A measurement of the concentration of a substance or pollutant within the immediate environs of an organism; taken to relate it to the amount of possible exposure.

Ambient Medium. Material surrounding or contacting an organism (*e.g.* outdoor air, indoor air, water, or soil, through which chemicals or pollutants can reach the organism.

Ambient Temperature. Temperature of the surrounding air or other medium.

Anadromous. Fish that spend their adult lives in the sea but swim upriver to freshwater spawning grounds to reproduce.

Anaerobe. An organism that lives and reproduces in the absence of dissolved oxygen, instead deriving oxygen from the breakdown of complex substances.

Anaerobic. A life or process that occurs in, or is not destroyed by, the absence of oxygen.

Anaerobic Decomposition. The breakdown of molecules into simpler molecules or atoms by microorganisms that can survive in the partial or complete absence of oxygen.

Anaerobic Digestion. Digestion of suspended organic matter by anaerobic microbial action.

Anaerobic Lagoon. A liquid-based manure management system, characterized by waste residing in water to a depth of at least six feet for a period ranging between 30 and 200 days. Bacteria produce methane in the absence of oxygen while breaking down waste.

Anaerobic Organism. An organism that does not need oxygen to stay alive.

Anaerobic Respiration. Metabolic process whereby electrons are transferred from an organic or in some cases, inorganic compounds to an inorganic acceptor molecule other than oxygen. The most common acceptors are nitrate, sulphate and carbonate.

Analog (climate). A large-scale *weather* pattern of the past that is similar to a current situation in its essential characteristics.

Ancillary Benefits. Complementary benefits of a climate policy including improvements in local air quality and reduced reliance of imported fossil fuels.

Anechoic. An environment that is very close to a free field.

Anhydrous. Very dry. No water or dampness is present.

Animal Dander. Tiny scales of animal skin, a common indoor air pollutant.

Animal Studies. Investigations using animals as surrogates for humans with the expectation that the results are pertinent to humans.

Anion Exchange Capacity. Sum total of exchangeable anions that a soil can adsorb. Expressed as centimoles of negative charge per kilogram of soil.

Annex I Parties. Industrialized countries that, as parties to the Framework Convention on Climate Change, have pledged to reduce their greenhouse gas emissions by the year 2000 to 1990 levels. Annex I Parties consist of countries belonging to the Organization for Economic Cooperation and Development (OECD) and countries designated as Economies-in-Transition. The 40 countries plus the European Economic Community listed in Annex I of the UNFCCC that agreed to try to limit their GHG emissions. Australia, Austria, Belarus, Belgium, Bulgaria, Canada, Croatia, Czech Republic, Denmark, European Economic Community, Estonia, Finland, France, Germany, Greece, Hungary, Iceland, Italy, Japan, Latvia, Liechtenstein, Lithuania, Luxembourg, Monaco, The Netherlands, New Zealand, Norway, Poland, Portugal, Romania, Russian Federation, Slovakia, Slovenia, Spain, Sweden, Switzerland, Turkey, Ukraine, United States.

Annex A. A list in the Kyoto Protocol of the six greenhouse gases and the sources of emissions covered under the Kyoto Protocol.

Annex B. A list in the Kyoto Protocol of 38 countries plus the European Community that agreed to QELRCs (emission targets), along with the QELRCs they accepted. The list is nearly identical to the Annex I Parties listed in the Convention except that it does not include Belarus or Turkey.

Antagonism. Interference or inhibition of the effect of one chemical by the action of another.

Antarctic Ice Sheet. See *ice sheet.*

Antarctic "Ozone Hole". Refers to the seasonal depletion of stratospheric ozone in a large area over Antarctica.

Anthracite. A hard, black, lustrous coal containing a high percentage of fixed carbon and a low percentage of volatile matter. Often referred to as hard coal.

Anthropogenic. Human made. In the context of greenhouse gases, emissions that are produced as the result of human activities.

Anti-Degradation Clause. Part of federal air quality and water quality requirements prohibiting deterioration where pollution levels are above the legal limit.

Anti-Microbial. An agent that kills microbes.

Anticyclone (high-pressure area). An atmospheric high-pressure closed circulation with clockwise rotation in the Northern Hemisphere, counterclockwise in the Southern Hemisphere, and undefined at the Equator.

Aquaculture. Production of food from aquatic habitats.

Aqueous. Something made up of water.

Aquifer. An underground geological formation, or group of formations, containing water. Aquifers are sources of groundwater for wells and springs.

Aquitard. Geological formation that may contain groundwater but is not capable of transmitting significant quantities of it under normal hydraulic gradients. May function as confining bed.

Arable Land. Land that can be cultivated to grow crops.

Aragonite. A mineral species of calcium carbonate ($CaCO_3$) with a crystal structure different from that of vaterite and calcite, which are the other two forms of $CaCO_3$. It is precipitated from ocean surface waters mainly by organisms (*e.g.* coral) that use it to make their shells and skeletons.

Arctic Haze. A persistent winter diffuse layer in the Arctic atmosphere whose origin may be related to long-range transport of midlatitude continental man-made pollutants.

Area Source. Any source of air pollution that is released over a relatively small area but which cannot be classified as a point source. Such sources may include vehicles and other small engines, small businesses and household activities, or biogenic sources such as a forest that releases hydrocarbons.

Aromatic. Applied to a group of hydrocarbons and their derivatives characterized by the presence of the benzene ring.

Aromatics. A type of hydrocarbon, such as benzene or toluene, with a specific type of ring structure. Aromatics are sometimes added to gasoline in order to increase octane. Some aromatics are toxic.

Arsenicals. Pesticides containing arsenic.

Artesian (Aquifer or Well). Water held under pressure in porous rock or soil confined by impermeable geological formations.

Asbestos. A mineral fiber that can pollute air or water and cause cancer or asbestosis when inhaled.

Ash. The mineral content of a product remaining after complete combustion.

Asphalt. A dark-brown-to-black cement-like material containing bitumen as the predominant constituent. It is obtained by petroleum processing. The definition includes crude asphalt as well as the following finished products. cements, fluxes, the asphalt content of emulsions (exclusive of water), and petroleum distillates blended with asphalt to make cutback asphalt.

Assigned Amount. In the Kyoto Protocol, the permitted emissions, in CO_2 equivalents, during a commitment period. It is calculated using the Quantified Emission Limitation and Reduction Commitment (QELRC), together with rules specifying how and what emissions are to be counted.

Assimilation. The ability of a body of water to purify itself of pollutants.

Assimilative Capacity. The capacity of a natural body of water to receive waste-waters or toxic materials without deleterious effects and without damage to aquatic life or humans who consume the water.

Atmosphere. The mixture of gases surrounding the Earth. The Earth's atmosphere consists of about 79.1% nitrogen (by volume), 20.9% oxygen, 0.036% carbon dioxide and trace amounts of other gases. The atmosphere can be divided into a number of layers according to its mixing or chemical characteristics, generally determined by its thermal properties (temperature). The layer nearest the Earth is the troposphere, which reaches up to an altitude of about 8 km (about 5 miles) in the polar regions and up to 17 km (nearly 11 miles) above the equator. The stratosphere, which reaches to an altitude of about 50 km (31 miles) lies atop the troposphere. The mesosphere which extends up to 80-90 km is atop the stratosphere, and finally, the thermosphere, or ionosphere, gradually diminishes and forms a fuzzy border with outer space. There is relatively little mixing of gases between layers.

Atmosphere (An). A standard unit of pressure representing the pressure exerted by a 29.92-in. column of mercury at sea level at 45 degrees latitude and equal to 1000 g/cm^2.

Atmosphere (The). The envelope of air surrounding the Earth and bound to it by the Earth's gravitational attraction. Studies of the chemical properties, dynamic motions, and physical processes of this system constitute the field of meteorology.

Atmospheric Inversion. A condition in which warmer air is found above cooler air, restricting air circulation; often associated in urban areas with a pollution event.

Atmospheric Response Variables. Variables that reflect the response of the atmosphere to external forcing (*e.g.*, temperature, pressure, circulation, and precipitation).

Atmospheric Turbulence. A state of the flow of air in which apparently random irregularities occur in the air's instantaneous velocities, often producing major deformations of the flow.

Atmospheric Windows. The range of wavelengths at which water vapor; carbon dioxide, or other atmospheric gases only slightly absorb radiation. Atmospheric windows allow the Earth's radiation to escape into space unless clouds absorb the radiation.

Atomic Weight. The average weight (or mass) of all the isotopes of an element, as determined from the proportions in which they are present in a given element, compared with the mass of the 12 isotope of carbon (taken as precisely 12.000), that is the official international standard; measured in daltons.

Atoms. Minute particles that are the basic building blocks of all chemical elements and thus all matter.

Attainment Area. An area considered to have air quality as good as or better than the national ambient air quality standards as defined in the Clean Air Act. An area may be an attainment area for one pollutant and a non-attainment area for others.

Attrition. Wearing or grinding down of a substance by friction. Dust from such processes contributes to air pollution.

Autotroph. An organism that produces its own food from inorganic compounds and a source of energy. There are photoautotrophs (photosynthetic plants) and chemical autotrophs.

Aviation Gasoline. All special grades of gasoline for use in aviation reciprocating engines, as given in the American Society for Testing and Materials (ASTM) specification D 910. Includes all refinery products within the gasoline range that are to be marketed straight or in blends as aviation gasoline without further processing (any refinery operation except mechanical blending). Also included are finished components in the gasoline range, which will be used for blending or compounding into aviation gasoline.

AWT. Advanced Waste Treatment—any process of water renovation that upgrades treated waste-water to meet reuse requirements.

B

Background Level. The concentration of a substance in an environmental media (air, water, or soil) that occurs naturally or is not the result of human activities.

Backwashing. Reversing the flow of water back through the filter media to remove entrapped solids.

Bacteria. One-celled organisms. Many act as decomposers that break down dead organic matter into substances that dissolve in water and are used as nutrients by plants.

Baghouse/Bag Filter. Fabric filter used for dust removal. Usually of tubular or envelope shape. The entire structure housing such bags is called a baghouse.

Ballistic Separator. A machine that sorts organic from inorganic matter for composting.

Baroclinic Model. A model of atmospheric circulation that, in contrast with barotropic models, does not constrain constant-pressure surfaces to coincide with constant-density surfaces.

Barrel (bbl). A liquid-volume measure equal to 42 United States gallons at 60 degrees Fahrenheit; used in expressing quantities of petroleum-based products.

Barrier Island. An island separated from the mainland by a salt marsh. It generally consists of a multiple system of beach ridges and is separated from other barrier islands by inlets that allow the exchange of seawater with lagoon water.

Basal Application. In pesticides, the application of a chemical on plant stems or tree trunks just above the soil line.

Basal Sliding (basal slip). The movement or speed of movement of a glacier on its bed.

Baseline Emissions. The emissions that would occur without policy intervention (in a business-as-usual scenario). Baseline estimates are needed to determine the effectiveness of emissions reduction programs (often called mitigation strategies).

Basic Solution. Water solution with more hydroxide ions (OH-) than hydrogen ions (H+); water solutions with pH greater than 7.

Basket of Gases. This refers to the group of six greenhouse gases regulated under the Kyoto Protocol. They are listed in Annex A of the Kyoto Protocol and include. carbon dioxide (CO_2), methane (CH_4), nitrous oxide (N_2O), hydrofluorocarbons (HFCs), perfluorocarbons (PFCs), and sulphur hexafluoride (SF_6).

Batch Process. A treatment process in which a tank or reactor is filled, the waste-water (or solution) is treated or a chemical solution is prepared and the tank is emptied. The tank may then be filled and the process repeated. Batch processes are also used to cleanse, stabilise or condition chemical solutions for use in industrial manufacturing and treatment processes.

Bathymetry. The science of measuring ocean depths to determine the topography of the sea floor.

Bay. A wide area of water extending into land from a sea or lake.

Becquerel. The unit commonly used for radioactive decay in the International System of measurement.

Benthic/Benthos. An organism that feeds on the sediment at the bottom of a water body such as an ocean, lake, or river.

Benthic Organism (benthos). A form of aquatic plant or animal life that is found on or near the bottom of a stream, lake, or ocean.

Benthic Region. The bottom layer of a body of water.

Berlin Mandate. A ruling negotiated at the first *Conference of the Parties* (COP 1), which took place in March, 1995, concluding that the present commitments under the *United Nations Framework Convention on Climate Change* are not adequate. Under the Framework Convention, developed countries pledged to take measures aimed at returning their greenhouse gas emissions to 1990 levels by the year 2000. The Berlin Mandate establishes a process that would enable the Parties to take appropriate action for the period beyond 2000, including a strengthening of developed country commitments, through the adoption of a protocol or other legal instruments.

Beta Particles. One of the three major kinds of nuclear radiation; electrons that are emitted when one of the protons or neutrons in the nucleus of an isotope spontaneously changes.

Bioaccumulants. Substances that increase in concentration in living organisms as they take in contaminated air, water, or food because the substances are very slowly metabolized or excreted.

Bioaccumulation. Accumulation of a chemical substance in living tissue.

Bioassay. A measurement of the effects of a substance on living organisms.

Biochemical Oxygen Demand (BOD). Amount of dissolved oxygen consumed in five days by biological processes breaking down organic matter.

Bioconcentration. The accumulation of a chemical in tissues of a fish or other organism to levels greater than in the surrounding medium.

Biodegradable. Material that can be broken down into simpler substances (elements and compounds) by bacteria or other decomposers. Paper and most organic wastes such as animal manure are biodegradable.

Biodegradation. Biological oxidation of natural or synthetic organic materials by soil micro-organisms, either in soils, water bodies, or waste-water treatment plants.

Biodiversity. Refers to the variety and variability among living organisms and the ecological complexes in which they occur. For biological diversity, these items are organized at many levels, ranging from complete ecosystems to the biochemical structures that are the molecular basis of heredity. Thus, the term encompasses different ecosystems, species, and genes.

Biofuel. Gas or liquid fuel made from plant material (biomass). Includes wood, wood waste, wood liquors, peat, railroad ties, wood sludge, spent sulfite liquors, agricultural waste, straw, tires, fish oils, tall oil, sludge waste, waste alcohol, municipal solid waste, landfill gases, other waste, and ethanol blended into motor gasoline.

Biogeochemical Cycle. The chemical interactions among the atmosphere, biosphere, hydrosphere, and lithosphere.

Biological Contaminants. Living organisms or derivates (*e.g.* viruses, bacteria, fungi, and mammal and bird antigens) that can cause harmful health effects when inhaled, swallowed, or otherwise taken into the body.

Biological Control. In pest control, the use of animals and organisms that eat or otherwise kill or out-compete pests.

Biological Evolution. The change in inherited characteristics of a population from generation to generation, which can result in new species.

Biological Integrity. The ability to support and maintain balanced, integrated, functionality in the natural habitat of a given region. Concept is applied primarily in drinking water management.

Biological Magnification. Refers to the process whereby certain substances such as pesticides or heavy metals move up the food chain, work their way into rivers or lakes, and are eaten by aquatic organisms such as fish, which in turn are eaten by large birds, animals or humans. The substances become concentrated in tissues or internal organs as they move up the chain.

Biological Oxidation. Decomposition of complex organic materials by microorganisms. Occurs in self-purification of water bodies and in activated sludge waste-water treatment.

Biological Oxygen Demand (BOD). Amount of dissolved oxygen needed by aerobic decomposers to break down the organic materials in a given volume of water at a certain temperature over a specified time period.

Biological Pesticides. Certain microorganism, including bacteria, fungi, viruses, and protozoa that are effective in controlling pests. These agents usually do not have toxic effects on animals and people and do not leave toxic or persistent chemical residues in the environment.

Biological Productivity. The amount of organic matter, carbon, or energy content that is accumulated during a given time period.

Biological Treatment. A treatment technology that uses bacteria to consume organic waste.

Biomagnification. Also called biological concentration. The tendency for some substances to concentrate with each trophic level. Organisms preferentially store certain chemicals and excrete others. When this occurs consistently among organisms, the stored chemicals increase as a percentage of the body weight as the material is transferred along a food chain or trophic level. For example, the concentration of DDT is greater in herbivores than in plants and greater in plants than in the non-living environment.

Biomass. Total dry weight of all living organisms that can be supported at each tropic level in a food chain. Also, materials that are biological in origin, including organic material (both living and dead) from above and below ground, for example, trees, crops, grasses, tree litter, roots, and animals and animal waste.

Biomass Energy. Energy produced by combusting biomass materials such as wood. The carbon dioxide emitted from burning biomass will not increase total atmospheric carbon dioxide if this consumption is done on a sustainable basis (*i.e.* if in a given period of time, regrowth of biomass takes up as much carbon dioxide as is released from biomass combustion). Biomass energy is often suggested as a replacement for fossil fuel combustion.

Biomass Fuel. A new name for the oldest fuel used by human. Organic matter, such as plant material and animal waste, that can be used as a fuel.

Biome. Entire community of living organisms in a single major ecological area.

Biomonitoring. 1. The use of living organisms to test the suitability of effluents for discharge into receiving waters and to test the quality of such waters downstream from the discharge. 2. Analysis of blood, urine, tissues, etc. to measure chemical exposure in humans.

Bioremediation. Use of living organisms to clean up oil spills or remove other pollutants from soil, water, or waste-water; use of organisms such as non-harmful insects to remove agricultural pests or counteract diseases of trees, plants, and garden soil.

Biosolid. The resides of waste-water treatment. Formerly called sewage sludge.

Biosphere. The portion of Earth and its *atmosphere* that can support life. The part (reservoir) of the global carbon cycle that includes living organisms (plants and animals) and life-derived organic matter (litter, detritus). The terrestrial biosphere includes the living biota (plants and animals) and the litter and soil organic matter on land, and the marine biosphere includes the biota and detritus in the oceans.

Biota. The animal and plant (fauna and flora) life of a given area.

Biotechnology. Techniques that use living organisms or parts of organisms to produce a variety of products (from medicines to industrial enzymes) to improve plants or animals or to develop microorganisms to remove toxics from bodies of water, or act as pesticides.

Biotic. Living: Living organisms make up the biotic parts of ecosystems.

Biotic Community. A naturally occurring assemblage of plants and animals that live in the same environment and are mutually sustaining and interdependent.

Bitumen. Gooey, black, high-sulfur, heavy oil extracted from tar sand and then upgraded to synthetic fuel oil.

Bituminous Coal. A dense, black, soft coal, often with well-defined bands of bright and dull material. The most common coal, with moisture content usually less than 20 per cent. Used for generating electricity, making coke, and space heating.

Black Carbon Aerosols. Particles of carbon in the atmosphere produced by inefficient combustion of fossil fuels or biomass. Black carbon aerosols absorb light from the sun, shading and cooling the Earth's surface, but contribute to significant warming of the atmosphere.

Blackwater. Water that contains animal, human, or food waste.

Blizzard. A severe weather condition characterized by low temperatures and strong winds (greater than 35 mph) bearing a great amount of snow, either falling or blowing. When these conditions persist after snow has stopped falling, it is called a ground blizzard.

Bloom. A proliferation of algae and/or higher aquatic plants in a body of water; often related to pollution, especially when pollutants accelerate growth.

BOD5. The biochemical oxygen demand of waste-water during decomposition occurring over a 5-day period. A measure of the organic content of waste-water.

Body Burden. The amount of concentration of a toxic chemical, especially radionuclides, in an individual.

Bog. A type of wetland that accumulates appreciable peat deposits. Bogs depend primarily on precipitacion for their water source, and are usually acidic and rich in plant residue with a conspicuous mat of living green moss.

Boreal. Of or relating to the forest areas of the northern North Temperate Zone, dominated by coniferous trees such as spruce, fir, and pine.

Borehole. Any exploratory hole drilled into the Earth or ice to gather geophysical data. Climate researchers often take *ice core* samples, a type of borehole, to predict atmospheric composition in earlier years.

Brackish. Mixed fresh and salt water.

Breathing Zone. Area of air in which an organism inhales.

Brine Mud. Waste material, often associated with well-drilling or mining, composed of mineral salts or other inorganic compounds.

British thermal unit (Btu). The quantity of heat required to raise the temperature of one pound of water one degree of Fahrenheit at or near 39.2 degrees Fahrenheit.

Bubble. An option in the Kyoto Protocol that allows a group of countries to meet their targets jointly by aggregating their total emissions. The member states of the European Union are utilizing this option.

Buffer Factor (Revelle factor). The ratio of the instantaneous fractional change in the partial pressure of CO_2 (pCO_2) exerted by seawater to the fractional change in total CO_2 dissolved in the ocean waters. The buffer factor relates the partial pressure of CO_2 in the ocean to the total ocean CO_2 concentration at constant temperature, alkalinity and salinity. The Revelle factor is a useful parameter for examining the distribution of CO_2 between the atmosphere and the ocean, and measures in part the amount of CO_2 that can be dissolved in the mixed surface layer.

Buffer Strips. Strips of grass or other erosion-resisting vegetation between or below cultivated strips or fields.

Buffers. Materials (chemicals) that have the ability to neutralise acids. Examples include the calcium carbonate that is present in many soils and rocks. These materials may lessen potential adverse effects of acid rain.

Bunker Fuel. Fuel supplied to ships and aircraft for international transportation, irrespective of the flag of the carrier, consisting primarily of residual and distillate fuel oil for ships and jet fuel for aircraft.

C

C_3 Plants. Plants (*e.g.*, soybean, wheat, and cotton) whose carbon-fixation products have three carbon atoms per molecule. Compared with C_4 plants, C_3 plants show a greater increase in photosynthesis with a doubling of CO_2 concentration and less decrease in stomatal conductance, which results in an increase in leaf-level water-use efficiency.

C_4 Plants. Plants (*e.g.*, maize and sorghum) whose carbon fixation products have four carbon atoms per molecule. Compared with C_3 plants, C_4 plants show little photosynthetic response to increased CO_2 concentrations above 340 ppm but show a decrease in stomatal conductance, which results in an increase in photosynthetic water-use efficiency.

Calcrete. A surficial gravel and sand conglomerate cemented by calcium carbonate.

Caliche. Also called hardpan; an opaque, reddish-brown-to-white calcareous material, which occurs in layers near the surface of stony soils in arid and semiarid areas.

Calvin Cycle. The incorporation of CO_2 into glucose by enzymatic reactions.

CAM Plants (crassulacean acid metabolism). Plants (*e.g.*, cactus and other succulents) that, unlike the C_3 and C_4 plants, temporarily separate the processes of carbon dioxide uptake and fixation when grown under arid conditions. They take up gaseous carbon dioxide at night when the stomata are open and water loss is minimal. During the day when the stomata are closed, the stored CO_2 is released and chemically processed. When CAM plants are not under water stress, they then follow C_3 photosynthesis.

Canal. A man-made watercourse designed to carry goods or water.

Canopy. The branches and leaves of woody plants that are formed some distance above the ground.

Canyon. A large but narrow gorge with deep sides.

Cap. A layer of clay, or other impermeable material installed over the top of a closed landfill to prevent entry of rainwater and minimize leachate.

Cape (or point). A piece of land extending into water.

Capillary Action. Movement of water through very small spaces due to molecular forces called capillary forces.

Capillary Fringe. The porous material just above the water table within which the porous medium is saturated by water under less than atmospheric pressure.

Carbon Adsorption. A treatment system that removes contaminants from groundwater or surface water by forcing it through tanks containing activated carbon treated to attract the contaminants.

Carbon-based Resources. The recoverable fossil fuel (coal, gas, crude oils, oil shale, and tar sands) and biomass that can be used in fuel production and consumption.

Carbon Black. An amorphous form of carbon, produced commercially by thermal or oxidative decomposition of hydrocarbons and used principally in rubber goods, pigments, and printer's ink.

Carbon Budget. The balance of the exchanges (incomes and losses) of carbon between the carbon reservoirs or between one specific loop (*e.g.* atmosphere—biosphere) of the carbon cycle. An examination of the carbon budget of a pool or reservoir can provide information about whether the pool or reservoir is functioning as a source or sink for CO_2.

Carbon Cycle. All parts (reservoirs) and fluxes of carbon; usually thought of as a series of the four main reservoirs of carbon interconnected by pathways of exchange. The four reservoirs, regions of the Earth in which carbon behaves in a systematic manner, are the atmosphere, terrestrial biosphere (usually includes fresh water systems), oceans, and sediments (includes fossil fuels). Each of these global reservoirs may be subdivided into smaller pools ranging in size from individual communities or ecosystems to the total of all living organisms (biota). Carbon exchanges from reservoir to reservoir by various chemical, physical, geological, and biological processes.

Carbon Density. The amount of carbon per unit area for a given ecosystem or vegetation type, based on climatic conditions, topography, vegetative-cover type and amount, soils, and maturity of the vegetative stands.

Carbon Dioxide (CO_2). A colorless, odorless, non-poisonous gas that is a normal part of the ambient air. Carbon dioxide is a product of fossil fuel combustion. Although carbon dioxide does not directly impair human health, it is a greenhouse gas that traps terrestrial (*i.e.* infrared) radiation and contributes to the potential for global warming.

Carbon Dioxide Equivalent (CDE). A metric measure used to compare the emissions from various greenhouse gases based upon their global warming potential (GWP). Carbon dioxide equivalents are commonly expressed as "million metric tons of carbon dioxide equivalents (MMTCDE)." The carbon dioxide equivalent for a gas is derived by multiplying the tons of the gas by the associated GWP. MMTCDE = (million metric tons of a gas) *(GWP of the gas)

Carbon Dioxide Fertilization. Enhancement of plant growth or of the net primary production by CO_2 enrichment that could occur in natural or agricultural systems as a result of an increase in the atmospheric concentration of CO_2.

Carbon Dioxide Reference Gas. A mixture of a known quantity of CO_2-in air or CO_2-in-N_2 used to calibrate carbon dioxide analyzers.

Carbon Equivalent (CE). A metric measure used to compare the emissions of the different greenhouse gases based upon their global warming potential (GWP). Greenhouse gas emissions in the U.S. are most commonly expressed as "million metric tons of carbon equivalents" (MMTCE). Global warming potentials are used to convert greenhouse gases to carbon dioxide equivalents—they can be converted to carbon equivalents by multiplying by 12/44 (the ratio of the molecular weight of carbon to carbon dioxide). The formula for carbon equivalents is.

MMTCE = (million metric tons of a gas) * (GWP of the gas) *(12/44)

Carbon Flux. The rate of exchange of carbon between pools (*i.e.* reservoirs).

Carbon Intensity. The relative amount of carbon emitted per unit of energy or fuels consumed.

Carbon Isotope Ratio. Ratio of carbon-12 to either of the other, less common, carbon isotopes, carbon-13 or carbon-14.

Carbon Pool. The reservoir containing carbon as a principal element in the geochemical cycle.

Carbon Sequestration. The uptake and storage of carbon. Trees and plants, for example, absorb carbon dioxide, release the oxygen and store the carbon. Fossil fuels were at one time biomass and continue to store the carbon until burned.

Carbon Sinks. Carbon reservoirs and conditions that take-in and store more carbon (*i.e.*, carbon sequestration) than they release. Carbon sinks can serve to partially offset greenhouse gas emissions. Forests and oceans are large carbon sinks.

Carbon Source. A pool (reservoir) that releases carbon to another part of the carbon cycle.

Carbon Taxes. A surcharge on the carbon content of oil, coal, and gas that discourages the use of fossil fuels and aims to reduce carbon dioxide emissions.

Carbon Tetrachloride (CCl_4). A compound consisting of one carbon atom and four chlorine atoms. It is an ozone depleting substance. Carbon tetrachloride was widely used as a raw material in many industrial applications, including the production of chlorofluorocarbons, and as a solvent. Solvent use was ended in the United States when it was discovered to be carcinogenic.

Carbonate-silicate Cycle. A complex biogeochemical cycle over time scales as long as one-half billion years. Included in this cycle are major geological processes, such as weathering, transport by ground and surface water, erosion and deposition of crustal rocks. The carbonate silicate cycle is believed to provide important negative feedback mechanisms that control the temperature of the atmosphere.

Carcinogen. Any material that is known to produce cancer in human or other animals.

Carnivores. Organisms that feed on other live organisms; usually applied to animals that eat other animals.

Carrying Capacity. The maximum abundance of a population or species that can be maintained by a habitat or ecosystem without degrading the ability of that habitat or ecosystem to maintain that abundance in the future.

Catanadromous. Fish that swim downstream to spawn.

Cation Exchange Capacity. The ability of a soil or other solid to exchange cations (positive ions such as calcium) with a liquid.

Cells. • In solid waste disposal, holes where waste is dumped, compacted, and covered with layers of dirt on a daily basis. • The smallest structural part of living matter capable of functioning as an independent unit.

Channelisation. An engineering technique that consists of straightening, deepening, widening, clearing or lining existing stream channels. The purpose is to control floods, improve drainage, control erosion, or improve navigation. It is a very controversial practice that may have significant environmental impacts.

Chemical Reaction. Interaction between chemicals in which there is a change in the chemical composition of the elements or compounds involved.

Chemoautotrophs. Autotrophic bacteria that can derive energy from chemical reactions of simple inorganic compounds. The most common use for inorganic sulphur compounds.

Chernozem (Tchernozem). A major group of dark-colored zonal soils with a rich and deep humus horizon occurring in temperate-to-cool, subhumid climates.

Chilling Effect. The lowering of the Earth's temperature because of increased particles in the air blocking the sun's rays.

Chisel Plowing. Preparing croplands by using a special implement that avoids complete inversion of the soil as in conventional plowing. Chisel plowing can leave a protective cover or crops residues on the soil surface to help prevent erosion and improve filtration.

Chlorofluorocarbons. A family of inert nontoxic and easily liquified chemicals used in refrigeration, air conditioning, packaging, and insulation or as solvents or aerosol propellants. Because they are not destroyed in the lower atmosphere, they drift into the upper atmosphere where their chlorine components destroy *ozone*.

Chloroplast. An organelle in the cells of green plants. It contains chlorophyll and functions in photosynthesis and protein synthesis.

Chlorosis. Discoloration of normally green plant parts caused by disease, lack of nutrients, or various air pollutants.

Clarification. A process in which suspended material is removed from a waste-water. This may be accomplished by sedimentation, with or without chemicals or filtration.

Clarified Waste-water. Waste-water that has had most of the settleable solids removed by clarification.

Clay. May refer to a mineral family or to a very finegrained sediment. It is associated with many environmental problems, such as shrinking and swelling of soils and sediment pollution.

Clean Development Mechanism (CDM). One of the three market mechanisms established by the Kyoto Protocol. The CDM is designed to promote sustainable development in developing countries and assist Annex I Parties in meeting their greenhouse gas emmissions reduction commitments. It enables industrialized countries to invest in emission reduction projects in developing countries and to receive credits for reductions achieved.

Clean Fuels. Blends or substitutes for gasoline fuels, including compressed natural gas, methanol, ethanol, and liquified petroleum gas.

Clear Cutting. A forest-management technique that involves harvesting all the trees in one area at one time.

Climate. The average weather, usually taken over a 30 year time period, for a particular region and time period. Climate is not the same as weather, but rather, it is the average pattern of weather for a particular region. Weather describes the short-term state of the atmosphere. Climatic elements include precipitation, temperature, humidity, sunshine, wind velocity, phenomena such as fog, frost, and hail-storms, and other measures of the weather.

Climate Change. The long-term fluctuations in temperature, precipitation, wind, and all other aspects of the Earth's climate. External processes, such as solar-irradiance variations, variations of the Earth's orbital parameters (eccentricity, precession, and inclination), lithosphere motions, and volcanic activity, are factors in climatic variation. Internal variations of the climate system also produce fluctuations of sufficient magnitude and variability to explain observed climate change through the feedback processes interrelating the components of the climate system.

Climate Feedback. An atmospheric, oceanic, terrestrial, or other process that is activated by direct climate change induced by changes in radiative forcing. Climate feedbacks may increase (positive feedback) or diminish (negative feedback) the magnitude of the direct climate change.

Climate Lag. The delay that occurs in climate change as a result of some factor that changes only very slowly. For example, the effects of releasing more carbon dioxide into the atmosphere may not be known for some time because a large fraction is dissolved in the ocean and only released to the atmosphere many years later.

Climate Model. A quantitative way of representing the interactions of the atmosphere, oceans, land surface, and ice. Models can range from relatively simple to quite comprehensive.

Climate Sensitivity. The magnitude of a climatic response to a perturbing influence. In mathematical modeling of the climate, the difference between simulations as a function of change in a given parameter.

Climate Signal. A statistically significant difference between the control and disturbed simulations of a climate model.

Climate System (or Earth system). The atmosphere, the oceans, the biosphere, the cryosphere, and the geosphere, together make up the climate system.

Climate Variability. Refers to changes in patterns, such as precipitation patterns, in the weather and climate.

Climate Variation. The change in one or more climatic variables over a specified time.

Climatic Analog. A past climate situation in which changes similar to the present occurred. Used in making climatic projections.

Climatic Anomaly. The deviation of a particular climatic variable from the mean or normal over a specified time.

Climatic Optimum. The period in history from about 5000 to about 2500 B.C. during which surface air temperatures were warmer than at present in nearly all regions of the world. In the Arctic region, the temperature rose many degrees, and in temperate regions, the increase was 1.0 degrees—1.7 degrees C. In this period, glaciers and ice sheets receded greatly, and the melt-water raised sea level by about 3 meters.

Climatology. Science dealing with *climate* and climate phenomena.

Closed-canopy Forest. Forests in which the leaves of adjacent trees overlap or touch, so that the trees form essentially continuous cover.

Cloud. A visible mass of condensed water vapor particles or ice suspended above the Earth's surface. Clouds may be classified on their visible appearance, height, or form.

Cloud Albedo. Reflectivity that varies from less than 10 to more than 90% of the insolation and depends on drop sizes, liquid water content, water vapor content, thickness of the cloud, and the sun's zenith angle. The smaller the drops and the greater the liquid water content, the greater the cloud albedo, if all other factors are the same.

Cloud Feedback. The coupling between cloudiness and surface air temperature in which a change in surface temperature could lead to a change in clouds, which could then amplify or diminish the initial temperature perturbation. For example, an increase in surface air temperature could increase the evaporation; this in turn might increase the extent of cloud cover. Increased cloud cover would reduce the solar radiation reaching the Earth's surface, thereby lowering the surface temperature. This is an example of negative feedback and does not include the effects of *longwave radiation* or the advection in the oceans and the atmosphere, which must also be considered in the overall relationship of the climate system.

Coagulants. Chemicals which cause very fine particles to clump (floc) together into larger particles. This makes it easier to separate the solids from the water by settling, skimming, draining or filtering.

Coagulation. In water or waste-water treatment, the destabilisation and initial aggregation of colloidal and finely divided suspended solids by the addition of floc-forming chemicals.

Coal. A black or brownish black solid, combustible substance formed by the partial decomposition of vegetable matter without access to air. The rank of coal, which includes anthracite, bituminous coal, subbituminous coal, and lignite, is based on fixed carbon, volatile matter, and heating value. Coal rank indicates the progressive alteration, or coalification, from lignite to anthracite.

Coal Coke. A hard, porous product made from baking bituminous coal in ovens at temperatures as high as 2,000 degrees Fahrenheit. It is used both as a fuel and as a reducing agent in smelting iron ore in a blast furnace.

Coal Gasification. Conversion of solid coal to synthetic natural gas (SNG) or a gaseous mixture that can be burned as a fuel.

Coal Liquefaction. Conversion of solid coal to a liquid fuel such as synthetic crude oil or methanol.

Coalbed Methane. Methane that is produced from coalbeds in the same manner as natural gas produced from other strata. Methane is the principal component of natural gas.

Coastal Zone. Lands and waters adjacent to the coast that exert an influence on the uses of the sea and its ecology or whose uses and ecology are affected by the sea.

Co-control Benefit. The additional benefit derived from an environmental policy that is designed to control one type of pollution, while reducing the emissions of other pollutants as well. For example, a policy to reduce carbon dioxide emissions might reduce the combustion of coal, but when coal combustion is reduced, so too are the emissions of particulates and sulfur dioxide. The benefits associated with reductions in emissions of particulates and sulfur dioxide are the co-control benefits of reductions in carbon dioxide.

Cogeneration. Production of two useful forms of energy such as high-temperature heat and electricity from the same process. For example, while boiling water to generate electricity, the leftover steam can be sold for industrial processes or space heating.

Combined Sewer. A sewer designed to carry both sanitary waste-water and storm or surface-water runoff.

Combustion. Chemical oxidation accompanied by the generation of light and heat.

Commensalism. A relationship between two kinds of organisms in which one benefits from the relationship and the other is neither helped nor hurt.

Commercial Waste. All solid waste emanating from business establishments such as stores, markets, office buildings, restaurants, shopping centers, and theaters.

Commitment Period. The period under the Kyoto Protocol during which Annex I Parties' GHG emissions, averaged over the period, must be within their emission targets. The first commitment period runs from January 1, 2008 to December 31, 2012.

Community, Ecological. A group of population of different species living in the same local area and interacting with one another. A community is the living portion of an ecosystem.

Composting. The natural biological *decomposition* of organic material in the presence of air to form a *humus*-like material.

Compound. Combination of two or more different chemical elements held together by chemical bonds.

Compressed Natural Gas (CNG). An alternative fuel for motor vehicles; considered one of the cleanest because of low hydrocarbon emissions and its vapors are relatively non-ozone producing. However, vehicles fueled with CNG do emit a significant quantity of nitrogen oxides.

Concentration. Amount of a chemical in a particular volume or weight of air, water, soil, or other medium. See parts per billion, parts per million.

Cone of Depression. A cone-shaped depression in the water table caused by withdrawal of water at rates greater than the rates at which the water can be replenished by natural groundwater flow.

Conference of the Parties (COP). The supreme body of the United Nations Framework Convention on Climate Change

(UNFCCC). It comprises more than 170 nations that have ratified the Convention. Its first session was held in Berlin, Germany, in 1995 and it is expected to continue meeting on a yearly basis. The COP's role is to promote and review the implementation of the Convention. It will periodically review existing commitments in light of the Convention's objective, new scientific findings, and the effectiveness of national climate change programs.

Conifer. See *coniferous trees*.

Coniferous Trees. Cone-bearing trees, mostly evergreens, that have needle-shaped or scale-like leaves. They produce wood known commercially as softwood.

Conservation. Preserving and renewing, when possible, human and natural resources. The use, protection, and improvement of natural resources according to principles that will ensure their highest economic or social benefits.

Consumptive Use. A type of off-stream water use. This water is consumed by plants and animals or in industrial processes and enters human tissue or products or evaporates during use. It is not returned to its source.

Contaminant. Any physical, chemical, biological, or radiological substance or matter that has an adverse effect on air, water, or soil.

Contamination. Presence of undesirable material that makes something unfit for a particular use.

Continental Crust. The layer of the Earth that lies under continents and the *continental shelves*. It ranges in thickness from 35 to 60 km. Its upper layer has a density of 2.7 g/cm^3 and is composed of rocks that are rich in silica and alumina.

Continental Plate. A thick continental crust.

Continental Shelves. Those parts of the continent that are covered by water. They are several to more than 322 km wide and about 122 m deep. At the edges of the shelves, the *continental slopes* drop rapidly from 100 to 200 m to 3000 to 3700 m.

Continental Slopes. See *continental shelves.*

Contour Plowing. Soil tilling method that follows the shape of the land to discourage erosion.

Controlled Burning. Using prescribed fire to reduce the risk from wildfires, control tree diseases, increase food and habitat for wildlife and manage forests for greater production of desirable tree species.

Convection. Atmospheric or oceanic motions that are predominately vertical and that result in vertical transport and mixing of atmospheric or oceanic properties. Because the most striking meteorological features result if atmospheric convective motion occurs in conjunction with the rising current of air (*i.e.* updrafts), convection is sometimes used to imply only upward vertical motion.

Convective Adjustment. A numerical procedure applied in many atmospheric models to approximate the vertical nonradiative heat transport. This procedure adjusts the lapse rate whenever necessary so that some prescribed critical lapse rate is never exceeded.

Conventional Pollutants. Statutorily listed pollutants understood well by scientists. These may be in the form of organic waste, sediment, acid, bacteria, viruses, nutrients, oil and grease, or heat.

Conventional Treatment. The preliminary treatment, sedimentation, floatation, trickling filter, rotating biological contactor, activated sludge and chlorination of wastewater.

Convergence. The quasi-horizontal flow of a fluid toward a common destination from different directions. When waters of different origins come together at a point or along a line (convergence line), the denser water from one side sinks under the lighter water from other side. The ocean convergence lines are the polar, subtropical, tropical, and equatorial.

Convergent Evolution. The process by which species evolve in different places or different times and although they have different genetic heritages, develop similar external forms and structures as a result of adaptation to similar environment. The similarity in the shapes of sharks and porpoises is an example of convergent evolution.

Convergent Plate Boundary. Boundary between two lithosphere plates in which one plate descends below the other (subduction).

Coriolis Effect. The tendency for an object moving above the Earth to turn to the right in the Northern Hemisphere and to the left in the Southern Hemisphere relative to the Earth's surface. The effect arises because the Earth rotates and is not, therefore, an inertial reference frame.

Cotyledon. A leaf or leaves of the embryos of seed plants. They can function in food storage and can become photosynthetic when the seed germinates.

Cover Crop. A crop that provides temporary protection for delicate seedlings and/or provides a cover canopy for seasonal soil protection and improvement between normal crop production periods.

Criteria Pollutant. A pollutant determined to be hazardous to human health and regulated under EPA's National Ambient Air Quality Standards. The 1970 amendments to the Clean Air Act require EPA to describe the health and welfare impacts of a pollutant as the "criteria" for inclusion in the regulatory regime. Emissions of the criteria pollutants CO, NO_x, NMVOCs, and SO_2 are thought to be precursors to greenhouse gas formation.

Crop Residue. Organic residue remaining after the harvesting and processing of a crop.

Crop Rotation. Planting the same field or areas of fields with different crops from year to year to reduce depletion of soil nutrients. A plant such as corn, tobacco, or cotton, which remove large amounts of nitrogen from the soil, is planted one year. The next year a legume such as soybeans, which add nitrogen to the soil, is planted.

Crop Water-use Efficiency. A measure at the ecosystem level of how well plants use available water in growth. The grams of dry weight gained by plants during the growing season per unit land area are divided by the millimeters of water lost (including evaporation directly from the soil).

Crude Oil. A mixture of hydrocarbons that exist in liquid phase in underground reservoirs and remain liquid at atmospheric pressure after passing through surface separating facilities.

Cryosphere. The portion of the climate system consisting of the world's ice masses and snow deposits, which includes the continental ice sheets, mountain glaciers, sea ice, surface snow cover, and lake and river ice. Changes in snow cover on the land surfaces are by and large seasonal and closely tied to the mechanics of atmospheric circulation. The glaciers and ice sheets are closely related to the global hydrologic cycle and to variations of sea level and change in volume and extent over periods ranging from hundreds to millions of years.

D

Deciduous Trees. Trees such as oaks and maples that lose their leaves during part of the year.

Decomposers. Heterotrophic organisms that break down dead protoplasm and use some of the products and release others for use by consumer organisms.

Decomposition. The breakdown of matter by bacteria and fungi. It changes the chemical composition and physical appearance of the materials.

Decomposition of Waste-Water. Breakdown of organic matter in waste-water by microbial action. It may be under aerobic or anaerobic conditions.

Decontamination. The removal or neutralisation of an infectious, poisonous or injurious agent from a site.

Deep Water. That part of the ocean below the main thermocline.

Deep-well Disposal. Method of disposal of hazardous liquid waste that involves pumping the waste deep into the ground below and completely isolated from all freshwater aquifers.

Deforestation. Those practices or processes that result in the change of forested lands to non-forest uses. This is often cited as one of the major causes of the enhanced greenhouse effect for two reasons: (1) the burning or decomposition of the wood releases carbon dioxide; and (2) trees that once removed carbon dioxide from the atmosphere in the process of photosynthesis are no longer present and contributing to carbon storage.

Degradable. See *biodegradable.*

Degradation. A growth phase in which the availability of food begins to limit cell growth.

Delta. The fan-shaped area at the mouth or lower end of a river; formed by eroded material that has been carried downstream and dropped in quantities larger than can be carried off by tides or currents.

Demineralisation. Removal of all salts from a water.

Dendrochronology. The dating of past events and variations in the environment and the climate by studying the annual growth rings of trees. The approximate age of a temperate forest tree can be determined by counting the annual growth rings in the lower part of the trunk. The width of these annual rings is indicative of the climatic conditions during the period of growth; wide annual rings signify favorable growing conditions, absence of diseases and pests, and favorable climatic conditions, while narrow rings indicate unfavorable growing conditions or climate.

Dendroclimatology. The use of tree growth rings as proxy climate indicators. Tree rings record responses to a wider range of climatic variables over a larger part of the Earth than any other type of annually dated proxy record.

Desalination. The removal of salts from seawater or brackish water so that the water can be used for purposes such as agriculture, industrial processes, or human consumption.

Desertification. The progressive destruction or degradation of vegetative cover especially in arid or semiarid regions bordering existing deserts. Overgrazing of rangelands, large-scale cutting of forests and woodlands, drought, and burning of extensive areas all serve to destroy or degrade the land cover. The climatic impacts of this destruction include increased albedo leading to decreased precipitation, which in turn leads to less vegetative cover; increased atmospheric dust loading could lead to decreased monsoon rainfall and greater wind erosion and/or atmospheric pollution.

Dew. Atmospheric moisture that condenses after a warm day and appears during the night on cool surfaces as small drops. The cool surfaces cause the water vapor in the air to cool to the point where the water vapor condenses.

Differentiation. The process by which single cells grow into particular forms of specialized tissue (*e.g.* root, stem, or leaf).

Disinfection. The killing or inactivation of most of the microorganisms in or on a substance with the probability that all pathogenic bacteria are killed by the disinfecting agent used.

Dispersion. A general term for a system consisting or particulate matter suspended in air or other gases.

Distillate Fuel Oil. A general classification for the petroleum fractions produced in conventional distillation operations. Included are products known as No. 1, No. 2, and No. 4 fuel oils and No. 1, No. 2, and No. 4 diesel fuels. Used primarily for space heating, on and off-highway diesel engine fuel (including railroad engine fuel and fuel for agricultural machinery), and electric power generation.

Divergence. A horizontal flow of water, in different directions, from a common center or zone; it is often associated with *upwelling*.

Divergent Evaluation. Organisms with the same ancestral genetic heritage migrate to different habitats and evolve into species with different external forms and structures, but typically continue to use the same kind of habitats.

Divergent Plate Boundary. Boundary between lithospheric plates characterised by the production of new lithosphere; found along oceanic ridges.

Diversity, Genetic. The total number of genetic characteristics, sometimes of a specific species, subspecies, or group of species.

Diversity, Habitat. The number of kinds of habitats in a given unit area.

Diversity, Species. Used loosely to mean the variety of species in an area or on the Earth. Technically, it is composed of three components: species richness—the total number of species; species evenness—the relative abundance of species; and species dominance—the most abundant species.

Dobson Unit. Commonly used to measure the concentration of ozone. One Dobson unit is equivalent to a concentration of 1 ppb ozone.

Doldrums. Region near the equator characterized by low pressure and light shifting winds.

Dominant Species. Generally, the species that are most abundant in an area, ecological community, or ecosystem.

Dominants. Species that are most abundant or otherwise most important within the community.

Downwelling. The process of accumulation and sinking of warm surface waters along a coastline. A change of air flow of the atmosphere can result in the sinking or downwelling of warm surface water. The resulting reduced nutrient supply near the surface affects the ocean productivity and meteorological conditions of the coastal regions in the downwelling area.

Drainage Basin. The area of land that drains water, sediment, and dissolved materials to a common outlet at some point along a stream channel.

Dredging. Removal of mud from the bottom of water bodies. This can disturb the ecosystem and causes silting that kills aquatic life. Dredging of contaminated muds can expose biota to heavy metals and other toxics.

Drip Irrigation. Irrigation by the application of water to the soil from tubes that drip water slowly, greatly reducing the loss of water from direct evaporation and increasing yield.

Dust. Finely divided particles formed by mechanical disintegration (size = 100 to several hundred microns). Dust particles usually are irregular in shape.

Dust Veil Index. A quantitative method developed by H.H. Lamb for comparing the magnitude of volcanic eruptions. The formulae use observations either of the depletion of the solar beam, temperature lowering in middle latitudes, or the quantity of solid matter dispersed as dust. The reference dust veil index is 1000, assigned to the Krakatoa 1883 eruption, and the index is calculated using all three methods, where the information is available, for statistical comparison purposes. Abbreviated D.V.I.

E

Early Crediting. A provision that allows crediting of emission reductions achieved prior to the start of a legally imposed emission control period. These credits can then be used to assist in achieving compliance once a legally imposed system begins.

Earth System Science. Science of studying the Earth as a system. Includes understanding of processes and linkages between the lithosphere, hydrosphere, biosphere and atmosphere.

Ecological Community. This term has two meanings. (i) A conceptual or functional meaning: a set of interacting species that occur in the same place (sometimes extended to mean a set that interacts in a way to sustain life); and (ii) An operational meaning: a set of species found in an area, whether or not they are interacting.

Ecological/Environmental Sustainability. Maintenance of ecosystem components and functions for future generations.

Ecological Gradient. A change in the relative abundance of a species or group of species along a line or over an area.

Ecological Impact. The effect that a man-caused or natural activity has on living organisms and their non-living (abiotic) environment.

Ecological Island. An area that is biologically isolated so that a species occurring within the area cannot mix (or only rarely mixes) with any other population of the same species.

Ecological Justification for the Conservation of Nature. An argument for the conservation of nature on the grounds that a species, an ecological community, an ecosystem or the Earth's biosphere provides specific functions necessary to the persistence of our life or of benefit to life. The ability of trees in forests to remove carbon dioxide produced in burning fossil fuels is such a public benefit and an argument for maintaining large areas of forests.

Ecological Niche. The general concept is that the niche is a species' 'profession'—what it does to make a living. The term is also used to refer to a set of environmental conditions within which a species is able to persist.

Ecological Succession. The process of the development of an ecological community or ecosystem, usually viewed as a series of stages—early, middle, late, mature (or climax) and sometimes postclimax. Primary succession is an original establishment; secondary succession is a re-establishment.

Ecology. The study of all aspects of how organisms interact with each other and/or their environment.

Ecosystem. The complex system of plant, animal, fungal, and microorganism communities and their associated non-living environment interacting as an ecological unit. Ecosystems have no fixed boundaries; instead their parameters are set to the scientific, management, or policy question being examined. Depending upon the purpose of analysis, a single lake, a watershed, or an entire region could be considered an ecosystem.

Eddy. A circular movement of water or air that is formed where currents pass obstructions or between two adjacent currents that are flowing counter to each other.

Effluent. Waste-water—treated or untreated—that flows out of a treatment plant, sewer, or industrial outfall. Generally refers to wastes discharged into surface waters.

El Chicon. Active volcano 7300 ft (2225 m) high in Mexico. The last eruption was in 1983.

Electrons. Tiny particle moving around outside the nucleus of an atom. Each electron has one unit of negative charge (-) and almost no mass.

Element. Chemicals such as hydrogen (H), iron (Fe), sodium (Na), carbon (C), nitrogen (N), or oxygen (O), whose distinctly different atoms serve as the basic building blocks of all matter. There are 92 naturally occurring elements. Another 15 have been made in laboratories. Two or more elements combine to form compounds that make up most of the world's matter.

El Ninõ. A climatic phenomenon occurring irregularly, but generally every 3 to 5 years. El Ninõs often first become evident during the Christmas season (El Ninõ means Christ child) in the surface oceans of the eastern tropical Pacific Ocean. The phenomenon involves seasonal changes in the direction of the tropical winds over the Pacific and abnormally warm surface ocean temperatures. The changes in the tropics are most intense in the Pacific region, these changes can disrupt weather patterns throughout the tropics and can extend to higher latitudes, especially in Central and North America. The relationship between these events and global weather patterns are currently the subject of much research in order to enhance prediction of seasonal to interannual fluctuations in the climate.

Emission. Materials (gases, particles, vapors, chemical compounds, etc.) that come out of smokestacks, chimneys, and tailpipes.

Emission Inventory. A list of air pollutants emitted into a community's, state's, nation's, or the Earth's atmosphere in amounts per some unit time (*e.g.* day or year) by type of source. An emission inventory has both political and scientific applications.

Emissions Cap. A mandated restraint in a scheduled time frame that puts a "ceiling" on the total amount of anthropogenic greenhouse gas emissions that can be released into the atmosphere. This can be measured as gross emissions or as net emissions (emissions minus gases).

Emissions Coefficient/Factor. A unique value for scaling emissions to activity data in terms of a standard rate of emissions per unit of activity (*e.g.*, grams of carbon dioxide emitted per barrel of fossil fuel consumed).

Emissions Trading. A market mechanism that allows emitters (countries, companies or facilities) to buy emissions from or sell emissions to other emitters. Emissions trading is expected to bring down the costs of meeting emission targets by allowing those who can achieve reductions less expensively to sell excess reductions (*e.g.* reductions in excess of those required under some regulation) to those for whom achieving reductions is more costly.

Emissivity. The ratio of the radiation emitted by a surface to that emitted by a black body at the same temperature.

Endangered Species. Animals, birds, fish, plants, or other living organisms threatened with extinction by anthropogenic (man-caused) or other natural changes in their environment.

Endemic. Species characteristic of or prevalent in a particular or restricted locality or region.

Energy. The capacity for doing work as measured by the capability of doing work (potential energy) or the conversion of this capability to motion (kinetic energy). Energy has several forms, some of which are easily convertible and can be changed to another form useful for work. Most of the world's convertible energy comes from fossil fuels that are burned to produce heat that is then used as a transfer medium to mechanical or other means in order to accomplish tasks. In the United States, electrical energy is often measured in kilowatt-hours (kWh), while heat energy is often measured in British thermal units (Btu).

Energy Balance Models. An analytical technique to study the solar radiation incident on the Earth in which explicit calculations of atmospheric motions are omitted. In the zero-dimensional models, only the incoming and outgoing radiation is considered. The outgoing *infrared radiation* is a linear function of global mean surface air temperature, and the reflected solar radiation is dependent on the surface albedo. The albedo is a step function of the global meansurface air temperatures, and equilibrium temperatures are computed for a range of values of the solar constant. The one-dimensional models have surface air temperature as a function of latitude. At each latitude, a balance between incoming and outgoing radiation and horizontal transport of heat is computed. (Abbreviated as EBM.)

Energy Conservation. Reduction or elimination of unnecessary energy use and waste.

Energy-efficiency. The ratio of the useful output of services from an article of industrial equipment to the energy use by such an article; for example, vehicle miles traveled per gallon of fuel (mpg).

Energy Intensity. Ratio between the consumption of energy to a given quantity of output; usually refers to the amount of primary or final energy consumed per unit of gross domestic product.

Energy Quality. Ability of a form of energy to do useful work. High-temperature heat and the chemical energy in fossil fuels and nuclear fuels are concentrated high quality energy. Low-quality energy such as low-temperature heat is dispersed or diluted and cannot do much useful work.

Energy Resources. The available supply and price of fossil and alternative resources will play a huge role in estimating how much a greenhouse gas constraint will cost.

Enhanced Greenhouse Effect. The concept that the natural greenhouse effect has been enhanced by anthropogenic emissions of greenhouse gases. Increased concentrations of carbon dioxide, methane, and nitrous oxide, CFCs, HFCs, PFCs, SF_6, NF_3, and other photochemically important gases caused by human activities such as fossil fuel consumption, trap more infra-red radiation, thereby exerting a warming influence on the climate.

Enhanced Oil Recovery. Removal of some of the heavy oil left in an oil well after primary and secondary recovery.

Enteric Fermentation. A digestive process by which carbohydrates are broken down by microorganisms into simple molecules for absorption into the bloodstream of an animal.

Environment. All external conditions that affect an organism or other specified system during its lifetime.

Environmental/Ecological Risk. The potential for adverse effects on living organisms associated with pollution of the environment by effluents, emissions, wastes, or accidental chemical releases; energy use; or the depletion of natural resources.

Environmental Equity/Justice. Equal protection from environmental hazards for individuals, groups, or communities regardless of race, ethnicity, or economic status. This applies to the development, implementation, and enforcement of environmental laws, regulations, and policies, and implies that no population of people should be forced to shoulder a disproportionate share of negative environmental impacts of pollution or environmental hazard due to a lack of political or economic strength levels.

Environmental Unity. A principle of environmental sciences that states that everything affects everything else, meaning that a particular course of action leads to an entire potential string of events.

Environmentalism. A social, political and ethical movement concerned with protecting the environment and using its resources wisely.

Epidemic Disease. A disease that appears occasionally in the population, affects a large percentage of it and declines or almost disappears for a while only to reappear later.

Epidemiology. The study of diseases as they affect populations, including the distribution of disease or other health-related states and events in human populations, the factors (*e.g.*, age, sex, occupation, and economic status) that influence this distribution, and the application of this study to control health problems.

Equilibrium Line. The level on a glacier where accumulation equals ablation and the net balance equals zero.

Erosion. The wearing away of land surface by wind or water, intensified by land-clearing practices related to farming, residential or industrial development, road building, or logging.

Estuary. Regions of interaction between rivers and near-shore ocean waters, where tidal action and river flow create a mixing of fresh and salt water.

Ethanol (C_2H_5OH). Otherwise known as ethyl alcohol, alcohol, or grain spirit. A clear, colorless, flammable oxygenated hydrocarbon with a boiling point of 78.5 degrees Celsius in the anhydrous state. In transportation, ethanol is used as a vehicle fuel by itself (E100), blended with gasoline (E85), or as a gasoline octane enhancer and oxygenate (10 per cent concentration).

Eukaryote. Organism whose cells have nuclei and certain other characteristics that separate it from the prokaryotes. The eukaryotes include flowering plants, animals and many single-cell organisms.

Euphotic Zone. The layer of a body of water that receives sufficient sunlight for photosynthesis. The depth of this layer, which is about 80 m, is determined by the water's extinction coefficient, the cloudiness, and the sunlight's angle of incidence.

Eutrophic. Referring to bodies of water having an abundance of the chemical elements required for life.

Eutrophication. The slow aging process during which a lake, estuary, or bay evolves into a bog or marsh and eventually disappears. During the later stages of eutrophication the water body is choked by abundant plant life due to higher levels of nutritive compounds such as nitrogen and phosphorus. Human activities can accelerate the process.

Evapotranspiration. Discharge of water from the Earth's surface to the atmosphere by evaporation from bodies of water, or other surfaces, and by transpiration from plants.

Even-aged Stands. Forest area where all live trees began growth from seeds and roots planted in about the same year.

Evolution Biological. The change in inherited characteristics of a population from generation to generation, sometimes resulting in a new species or population.

Exosphere. The uppermost layer of the atmosphere, its lower boundary is estimated at 500 km to 1000 km above the Earth's surface. It is only from the exosphere that atmospheric gases can, to any appreciable extent, escape into outer space.

Exponential Growth. Growth in which some quantity, such as population size, increases by a constant percentage of the whole during each year or other time period; when the increase in quantity over time is plotted, this type of growth yields a curve shaped like the letter J.

F

Facultative Anaerobe. A bacterium capable of growing under aerobic conditions or anaerobic conditions in the presence of an inorganic ion, *i.e.*, SO_4, NO_3.

Facultative Stabilisation Pond. Basin for retention of a wastewater, in which biological oxidation of the organic matter occurs by bacterial action. A significant amount of molecular oxygen used by the microbes is supplied by algal photosynthesis.

Fall Line. The point on a river where there is an abrupt drop in elevation of the land and where numerous waterfalls occur. The line is typically located where streams pass from harder to softer rocks.

Fallow. A field allowed to grow with a cover crop without harvesting for at least one season.

Fecal Coliform Bacteria. Standard measure of microbial pollution and an indicator of disease potential for a water source.

Feedback. A kind of system response that occurs when output of the system also serves as input leading to changes in the system.

Feedback Mechanisms. A mechanism that connects one aspect of a system to another. The connection can be either amplifying (positive feedback) or moderating (negative feedback).

Feedlot. Confined outdoor or indoor space used to raise hundreds to thousands of domesticated livestock.

Fen. A type of *wetland* that accumulates peat deposits; they are less acidic than bogs, deriving most of their water from groundwater rich in calcium and magnesium.

Fertilization. A term used to denote efforts to enhance plant growth by increased application of nitrogen-based fertilizer or increased deposition of nitrates in precipitation.

Fertilization, Carbon Dioxide. An expression (sometimes reduced to *fertilization*) used to denote increased plant growth due to a higher carbon dioxide concentration.

Fertilizer. Substance that adds inorganic or organic plant nutrients to soil and improves its ability to grow crops, trees, or other vegetation.

Fill. Man-made deposits of natural soils or rock products and waste materials.

Firn. Material that is transitional between snow and glacier ice. It is formed from snow after passing through one summer melt season and becomes glacier ice after its permeability to liquid water falls to zero.

First Detection. Identification of a precursor signal, detectable above the noise of natural climatic variability, of a significant change in a climate parameter and attribution of this change to an increase in atmospheric carbon dioxide concentration. The signal may be estimated by numeric modeling of the climate, and the noise can be estimated using instrumental data. For any modeled signal that is estimated, the corresponding noise can be estimated from observational data, and a signal-to-noise ratio can be calculated to provide a quantitative measure of detectability.

Flaring. The burning of waste gases through a flare stack or other device before releasing them to the air.

Flow Law. In glaciology, a constitutive relation for the analysis of three-dimensional deformation states of ice subjected to stress.

Flue Gas. Gaseous product of combustion, for example, a boiler furnace or a kiln.

Fluidized Bed Combustion (FBC). Process for burning coal more efficiently, cleanly, and cheaply. A stream of hot air is used to suspend a mixture of powdered coal and limestone during combustion. About 90 to 98 per cent of the sulfur dioxide produced during combustion is removed by reaction with limestone to produce solid calcium sulfate.

Fluorocarbons. Carbon-fluorine compounds that often contain other elements such as hydrogen, chlorine, or bromine. Common fluorocarbons include chlorofluorocarbons (CFCs), hydrochlorofluorocarbons (HCFCs), hydrofluorocarbons (HFCs), and perfluorocarbons (PFCs).

Fly Ash. Finely divided particles of ash entrained in flue gases arising from the combustion of fossil fuels (predominantly coal). The particles of ash may contain incompletely burned fuel.

Fog. Liquid particles less than 40 microns in diameter that are formed by condensation of *vapor* in air.

Food Chain. A sequence of organisms, each of which uses the next lower member of the sequence as a food source.

Forcing Mechanism. A process that alters the energy balance of the climate system, *i.e.* changes the relative balance between incoming solar radiation and outgoing infrared radiation from Earth. Such mechanisms include changes in solar irradiance, volcanic eruptions, and enhancement of the natural greenhouse effect by emission of carbon dioxide.

Forest. Terrestrial ecosystem (biome) with enough average annual precipitation (at least 76 centimeters or 30 inches) to support growth of various species of trees and smaller forms of vegetation.

Fossil Fuel. A general term for buried combustible geologic deposits of organic materials, formed from decayed plants and animals that have been converted to crude oil, coal, natural gas, or heavy oils by exposure to heat and pressure in the earth's crust over hundreds of millions of years.

Fossil Fuel Combustion. Burning of coal, oil (including gasoline), or natural gas. This burning, usually to generate energy, releases carbon dioxide, as well as combustion by products that can include unburned hydrocarbons, methane, and carbon monoxide. Carbon monoxide, methane, and many of the unburned hydrocarbons slowly oxidize into carbon dioxide in the atmosphere. Common sources of fossil fuel combustion include cars and electric utilities.

Freon. See *chlorofluorocarbons*.

Fugitive Emissions. Unintended gas leaks from the processing, transmission, and/or transportation of fossil fuels, CFCs from refrigeration leaks, SF_6 from electrical power distributor, etc.

Fugitive Sources. Type of stationary air pollution sources that generate pollutants from open areas exposed to wind processes.

Fungi. Molds, mildews, yeasts, mushrooms, and puffballs, a group of organisms that lack chlorophyll and therefore are not photosynthetic. They are usually nonmobile, filamentous, and multicellular.

G

Gaia. The Gaia hypothesis states that the surface environment of the Earth, with respect to such factors as the atmospheric composition of reactive gases (for example, oxygen, carbon dioxide and methane), the acidity-alkalinity of water and the surface temperature are actively regulated by the sensing, growth, metabolism and other activities of the biota. Interaction between the physical and biological system on the Earth's surface has led to a planetwide physiology which began more than 3 billion years ago and the evolution of which can be detected in the fossil record.

Game Fish. Species like trout, salmon, or bass, caught for sport. Many of them show more sensitivity to environmental change than "rough" fish.

Game Ranching. Practice of maintaining wild herbivores in their native habitat to be harvested for meat, leather and other products.

Garbage. Animal and vegetable waste resulting from the handling, storage, sale, preparation, cooking, and serving of foods.

Gasohol. Vehicle fuel consisting of a mixture of gasoline and ethyl or methyl alcohol; typically 10 to 23 per cent ethanol by volume.

General Aviation. That portion of civil aviation, which encompasses all facets of aviation except air carriers. It includes any air taxis, commuter air carriers, and air travel clubs, which do not hold Certificates of Public Convenience and Necessity.

General Circulation Models. Hydrodynamic models of the atmosphere on a grid or spectral resolution that determine the surface pressure and the vertical distributions of velocity, temperature, density, and water vapor as functions of time from the mass conservation and hydrostatic laws, the first law of thermodynamics, Newton's second law of motion, the equation of state, and the conservation law for water vapor. Abbreviated as GCM. Atmospheric general circulation models are abbreviated AGCM, while oceanic general circulation models are abbreviated OGCM.

Genetic Drift. Changes in the frequency of a gene in a population as a result of chance rather than of mutation, selection or migration.

Genetic Risk. Used in discussions of endangered species to mean detrimental change in genetic characteristics not caused by external environmental changes. Genetic changes can occur in small population from such causes as reduced genetic variation, genetic drift and mutation.

Geographic Information System (GIS). Technology capable of storing, retrieving, transferring and displaying environmental data.

Geologic Cycle. The formation and destruction of earth materials and the processes responsible for these events. The geologic cycle includes the following subcycles: hydrologic, tectonic, rock and geochemical.

Geomorphology. The study of present-day landforms, including their classification, description, nature, origin, development, and relationships to underlying structures. Also the history of geologic changes as recorded by these surface features. The term is sometimes restricted to features produced only by erosion and deposition.

Geopressurised Systems. Geothermal system that exists when the normal heat flow from the Earth is trapped by impermeable clay layers that act as an effective insulator.

Geosphere. The solid mass (lithosphere) of the Earth as distinct from the atmosphere and hydrosphere or all three of these layers combined.

Geostrophic Flow. A type of movement where the Coriolis force balances exactly the horizontal pressure force.

Geothermal Energy. Heat transferred from the earth's molten core to under-ground deposits of dry steam (steam with no water droplets), wet steam (a mixture of steam and water droplets), hot water, or rocks lying fairly close to the earth's surface.

Giardia Lamblia. Protozoan in the feces of humans and animals that can cause severe gastrointestinal ailments. It is a common contaminant of surface waters.

Glacial Maximum. The position or time of the greatest advance of a glacier (*e.g.* the greatest equatorward advance of Pleistocene glaciation).

Glacial Rebound. The isostatic adjustment of previously glaciated areas after glacial retreat (*e.g.*, the uplift of Scandinavia after the most recent glaciation).

Glacier. A mass of land ice that is formed by the cumulative recrystallization of firn. A glacier flows slowly (at present or in the past) from an accumulation area to an ablation area. Some well-known glaciers are: the Zermatt, Stechelberg, Grindelwald, Trient, Les Diablerets, and Rhone in Switzerland; the Nigards, Gaupne, Fanarak, Lom, and Bover in Norway; the Wright, Taylor, and Wilson Piedmont glaciers in Antarctica; the Bossons Glacier in France; the Emmons and Nisqually glaciers on Mt. Ranier, Washington; Grinnell glacier in Glacier National Park, Montana; the Dinwoody glacier in the Wind River Mountains and the Teton glacier in Teton National Park, both in Wyoming; and many glaciers in the Canadian Rockies.

Glacier Flow (ice flow). The slow downward or outward movement of ice in a glacier caused by gravity.

Global Circulation Models (GCM). A type of mathematical model used to evaluate global change, particularly related to climatic change. GCMs are very complex and require supercomputers for their operation.

Global Extinction. Disappearance or extinction of a species everywhere.

Global Forecasting. Process of predicting or forecasting future change in environmental areas such as world population, natural resource utilisation and environmental degradation.

Global Warming. The progressive gradual rise of the earth's surface temperature thought to be caused by the greenhouse effect and responsible for changes in global climate patterns. An increase in the near surface temperature of the Earth. Global warming has occurred in the distant past as the result of natural influences, but the term is most often used to refer to the warming predicted to occur as a result of increased emissions of greenhouse gases.

Global Warming Potential (GWP). The index used to translate the level of emissions of various gases into a common measure in order to compare the relative radiative forcing of different gases without directly calculating the changes in atmospheric concentrations. GWPs are calculated as the ratio of the radiative forcing that would result from the emissions of one kilogram of a greenhouse gas to that from emission of one kilogram of carbon dioxide over a period of time (usually 100 years). The chart below shows the original GWPs (assigned in 1990) and the most recent GWPs (assigned in 1996) for the most important greenhouse gases.

GAS	GWP 1990	GWP 1996
Carbon Dioxide	1	1
Methane	22	21
Nitrous Oxide	270	310
HFC-134a	1,200	1,300
HFC-23	10,000	11,700
HFC-152a	150	140
HCF-125	NA*	2,800
PFCs**	5,400	7,850
SF6	NA*	23,900

* Not Applicable. GWP was not yet estimated for this gas.

**This figure is an average GWP for the two PFCs, CF_4 and C_2F_6.

See *radiative forcing, carbon equivalent, carbon dioxide equivalent.*

Grassland. Terrestrial ecosystem (biome) found in regions where moderate annual average precipitation (25 to 76 centimeters or 10 to 30 inches) is enough to support the growth of grass and small plants but not enough to support large stands of trees.

Gravel. Unconsolidated, generally rounded fragment of rocks and minerals greater than 2 mm in diameter.

Green Plans. Long-term strategies for identifying and solving global and regional environmental problems. The philosophical heart of green plants is sustainability.

Greenbelt. The idea of locating garden cities in a set connected by undeveloped areas, forming a system of countryside and urban landscapes.

Greenhouse Effect. The effect produced as greenhouse gases allow incoming solar radiation to pass through the Earth's atmosphere, but prevent part of the outgoing infrared radiation from the Earth's surface and lower atmosphere from escaping into outer space. This process occurs naturally and has kept the Earth's temperature about 59 degrees F warmer than it would otherwise be. Current life on Earth could not be sustained without the natural greenhouse effect.

Greenhouse Gas. Any gas that absorbs infra-red radiation in the atmosphere. Greenhouse gases include water vapor, carbon dioxide (CO_2), methane (CH_4), nitrous oxide (N_2O), halogenated fluorocarbons (HCFCs), ozone (O_3), perfluorinated carbons (PFCs), and hydrofluorocarbons (HFCs).

Greenland Ice Sheet. See *ice sheet*.

Grenz. A soil horizon, which is frequently marked by a bed of clay, that results from a temporary halt in the accumulation of vegetal material.

Grit. Dense, mineral, suspended matter present in a water or waste-water, such as silt and sand.

Grit Removal. Removal of heavy suspended mineral matter present in a waste-water, such as sand and silt.

Gross Primary Production. The total amount or weight of organic matter created by photosynthesis over a defined time period (total product of photosynthesis). Abbreviated GPP.

Ground Cover. Plants grown to keep soil from eroding.

Grounding Line. The boundary between the area where an ice shelf or a glacier is floating on water and where it is in contact with the shore or underlying earth (grounded).

Groundwater. The supply of fresh water found beneath the surface of the Earth (usually in aquifers) that often supplies wells and springs.

Growth Rate. The net increase in some factor per unit time. In ecology, the growth rate of a population is sometimes measured as the increase in numbers of individuals or biomass per unit time and sometimes as a percentage increase in numbers or biomass per unit time.

Growth Water-use Efficiency. A measure at the individual plant level of how well plants use available water in growth. The units of dry matter synthesized are divided by the units of water lost.

Gulf. A large arm of an ocean or sea extending into a land mass.

Gulf Stream. A warm, swift ocean current that flows along the coast of the Eastern United States and makes Ireland, Great Britain, and the Scandinavian countries warmer than they would be otherwise.

Gulf Stream Meander. A transient winding bend in the Gulf Stream. These bends intensify as the Gulf Stream merges into the North Atlantic and can break up into detached eddies at about 40 degrees S.

Gyres. Major circular flow patterns in the oceans. The wind-driven eastward- and westward-flowing equatorial currents are blocked by the continents and rotate slowly in a clockwise direction in the North Atlantic and Pacific Oceans and in a counter-clockwise direction in the South Atlantic, South Pacific, and Indian Oceans.

H

Habitat. The area or region where a particular type of plant or animal lives and grows.

Hadley Cell. A direct thermally-driven and zonally symmetric large-scale atmospheric circulation first proposed by George Hadley in 1735 as an explanation for the trade winds. It carries momentum, sensible heat, and potential heat from the tropics to the mid-latitudes (30 degrees). The poleward transport aloft is complemented by subsidence in the subtropical high pressure ridge and a surface return flow. The variability of this cell and the *Walker cell* is hypothesized to be a major factor in short-term climatic change.

Hail. Precipitation composed of balls or irregular lumps of ice. Hail is produced when large frozen raindrops, or almost any particles, in cumulonimbus clouds act as embryos that grow by accumulating supercooled liquid droplets. Violent updrafts in the cloud carry the particles in freezing air; allowing the frozen core to accumulate more ice. When the piece of hail becomes too heavy to be carried by upsurging air currents it falls to the ground.

Halocarbons. Chemicals consisting of carbon, sometimes hydrogen, and either chlorine, fluorine bromine or iodine.

Halocline. In the oceans, a well-defined vertical gradient of salinity.

Halons. Compounds, also known as bromofluorocarbons, that contain bromine, fluorine, and carbon. They are generally used as fire extinguishing agents and cause ozone depletion. Bromine is many times more effective at destroying stratospheric ozone than chlorine.

Hazardous Air Pollutants. Air pollutants which are not covered by ambient air quality standards but which may present a threat of adverse human health effects or adverse environmental effects. Such pollutants include asbestos, beryllium, mercury, benzene, coke oven emissions, radionuclides, and vinyl chloride.

Hazardous Waste Minimization. Reducing the amount of toxicity or waste produced by a facility via source reduction or environmentally sound recycling.

Haze. Fine dry or wet particles of dust, salt, or other impurities that can concentrate in a layer next to the Earth when air is stable.

Heat Content. The amount of heat per unit mass released upon complete combustion.

Heat Energy. Energy of the random motion of atoms and molecules.

Heat Flux (thermal flux). The amount of heat that is transferred across a surface of unit area in a unit of time.

Heat Island. Warmer air of a city than surrounding areas as a result of increased heat production and decreased rate of heat loss caused by the abundance of building and paving materials, which act as solar collectors.

Heat Island Effect. A dome of elevated temperatures over an urban area caused by the heat absorbed by structures and pavement.

Herbivore. An animal that feeds on plants.

Heterotrophs. Organisms that break down and use organic matter.

Heavy Metal. Refers to a number of metals, including lead, mercury, arsenic and silver (among others) that have a relatively high atomic number (the number of protons in the nucleus of an atom). They are often toxic at relatively low concentrations, causing a variety of environmental problems.

HGWP (High Global Warming Potential). Some industrially produced gases such as sulfur hexafluoride (SF_6), perfluorocarbons (PFCs), and hydrofluorocarbons (HFCs) have extremely high GWPs. Emissions of these gases have a much greater effect on global warming than an equal emmission (by weight) of the naturally occurring gases. Most of these gases have GWPs 1,300 - 23,900 times that of CO_2. These GWPs can be compared to the GWPs of CO_2, CH_4, and N_2O which are presently estimated to be 1, 23 and 296, respectively.

Higher Heating Value. Quantity of heat liberated by the complete combustion of a unit volume or weight of a fuel assuming that the produced water vapor is completely condensed and the heat is recovered; also known as gross calorific value.

Histosol. Wet organic soils, such as peats and mucks.

Holdridge Life Zone. A climate category defined by three weighted climatic indexes, namely, mean annual heat, precipitation, and atmospheric moisture.

Holocene. The most recent epoch of the *Quaternary period*, covering approximately the last 10,000 years.

Homeostasis. The ability of a cell or organism to maintain a constant environment. Results from negative feedback, resulting in a state of dynamic equilibrium.

"Hot Air". A situation in which emissions (of a country, sector, company or facility) are well below a target due to the target being above emissions that materialized under the normal course of events (*i.e.* without deliberate emission reduction efforts). Hot air can result from over-optimistic projections of growth. Emissions are often projected to grow roughly in proportion to GDP, and GDP is often projected to grow at historic rates. If a recession occurs and fuel use declines, emissions may be well below targets since targets are generally set in relation to emission projections. If emissions trading is allowed, an emitter could sell the difference between actual emissions and emission targets. Such emissions are considered hot air because they do not represent reductions from what would have occurred in the normal course of events.

Human Demography. The study of human population characteristics, such as age structure, demographic transition, total fertility, human population and environment relationships, death-rate factors and standard of living.

Humidity. The amount of water vapor in the air. The higher the temperature, the greater the number of water molecules the air can hold.

Humus. Decomposed organic material.

Hurricanes. Severe tropical storms whose winds exceed 74 mph. Hurricanes originate over the tropical and subtropical North Atlantic and North Pacific oceans, where there is high humidity and light wind.

Hutchinsonian Niche. The idea of a measured niche, a set of environmental conditions within which a species is able to persist.

Hydrocarbons. Substances containing only hydrogen and carbon. Fossil fuels are made up of hydrocarbons. Some hydrocarbon compounds are major air pollutants.

Hydrochlorofluorocarbons (HCFCs). Compounds containing hydrogen, fluorine, chlorine, and carbon atoms. Although ozone depleting substances, they are less potent at destroying stratospheric ozone than chlorofluorocarbons (CFCs). They have been introduced as temporary replacements for CFCs and are also greenhouse gases.

Hydroelectric Power Plant. Structure in which the energy of fading or flowing water spins a turbine generator to produce electricity.

Hydrofluorocarbons (HFCs). Compounds containing only hydrogen, fluorine, and carbon atoms. They were introduced as alternatives to ozone depleting substances in serving many industrial, commercial, and personal needs. HFCs are emitted as by-products of industrial processes and are also used in manufacturing. They do not significantly deplete the stratospheric ozone layer, but they are powerful greenhouse gases with global warming potentials ranging from 140 (HFC-152a) to 11,700 (HFC-23).

Hydrologic Budget. A quantitative accounting of all water volumes and their changes with time for a basin or area.

Hydrologic Cycle. The process of evaporation, vertical and horizontal transport of vapor, condensation, precipitation, and the flow of water from continents to oceans. It is a major factor in determining climate through its influence on surface vegetation, the clouds, snow and ice, and soil moisture. The hydrologic cycle is responsible for 25 to 30 per cent of the mid-latitudes' heat transport from the equatorial to polar regions.

Hydrology. The science dealing with the properties, distribution, and circulation of water.

Hydroponics. The practice of growing plants in a fertilised water solution on a completely artificial substrate in an artificial environment such as a greenhouse.

Hydropower. Electrical energy produced by falling or flowing water.

Hydrosphere. The part of the Earth composed of water including clouds, oceans, seas, ice caps, glaciers, lakes, rivers, underground water supplies, and atmospheric water vapor.

Hydrostatic Equation. In the vector equation of motion, the form assumed by the vertical component when all Coriolis, earth-curvature, frictional, and vertical-acceleration terms are considered negligible compared with those involving the vertical pressure force and the force of gravity. The error in applying the hydrostatic equation to the atmosphere for cyclonic-scale motions is less than 0.01%. In extreme situations, the strong vertical accelerations in thunderstorms and mountain waves can be 1% of gravity.

Hypolimnion. Bottom waters of a thermally stratified lake. The hypolimnion of a eutrophic lake is usually low or lacking in oxygen.

Hypsithermal Period. The period about 4000 to 8000 years ago when the Earth was apparently several degrees warmer than it is now. More rainfall occurred in most of the subtropical desert regions and less in the central midwest United States and Scandinavia. It is also called the altithermal period and can serve as a past climate analog for predicting the regional pattern of climate change should the mean Earth surface temperature increase from an increase in atmospheric carbon dioxide concentration.

I

Ice Age. A glacial epoch or time of extensive glacial activity. Also, as Ice Age, which refers to the latest glacial epoch, the *Pleistocene* Epoch.

Ice and Snow Albedo. The reflectivity of ice and snow-covered surfaces. The albedo of freshly fallen snow may be as much as 90%, while older snow may have values of 75% or less. The larger the areal extent of snow and ice cover, the higher the albedo value. The surface albedo will also increase as a function of the depth of snow cover up to 13 cm and be unaffected by increased snow cover after reaching that depth.

Ice and Snow-albedo-temperature Feedback. Interactions that can be described as a theoretical concept of a feedback mechanism in which the interacting elements are the areal extent of polar ice and snow cover, the albedo of the polar region (dependent on areal extent of ice and snow), absorption of solar radiation (dependent on the albedo), temperature (dependent on the absorption of solar radiation) and the area of ice and snow cover (dependent on temperature). Less snowfall would mean more absorption of solar radiation, therefore a surface warming would occur. Climate modeling studies indicate an amplification effect (*i.e. positive feedback*) of the ice and snow-albedo feedback on increased surface air temperatures caused by increases in the atmospheric concentration of carbon dioxide.

Ice Core. A cylindrical section of ice removed from a glacier or an ice sheet in order to study climate patterns of the past. By performing chemical analyses on the air trapped in the ice, scientists can estimate the percentage of carbon dioxide and other trace gases in the atmosphere at that time.

Ice Cover. During the present time, the extent, especially the thickness, of glacier ice on a land surface. Also the same as ice concentration, which is the ratio of an area of sea ice to the total area of sea surface within some large geographic area.

Ice Flow. See *glacier flow*.

Ice Front. The floating vertical cliff that forms the seaward face or edge of a *glacier* or an *ice shelf* that enters water. It can vary from 2 to 50 m in height.

Ice Sheet (continental glacier). A glacier of considerable thickness and more than 50,000 sq km in area. It forms a continuous cover of ice and snow over a land surface. An ice sheet is not confined by the underlying topography but spreads outward in all directions. During the *Pleistocene* Epoch, ice sheets covered large parts of North America and northern Europe but they are now confined to polar regions (*e.g.* Greenland and Antarctica).

Ice Shelf. A sheet of very thick ice with a level or gently undulating surface. It is attached to the land on one side, but most of it is floating. On the seaward side, it is bounded by a steep cliff (*ice front*) 2 to 50 m or more above sea level. Ice shelves have formed along polar coasts (*e.g.* Antarctica and Greenland); they are very wide with some extending several hundreds of kilometers toward the sea from the coastline. They increase in size from annual snow accumulation and seaward extension of land *glaciers*. They decrease in size from warming, melting, and calving.

Igneous Rocks. Rocks formed from the solidification of magma. They are extrusive if they crystallise on the surface of the Earth and intrusive if they crystallise beneath the surface.

Imminent Hazard. One that would likely result in unreasonable adverse effects on humans or the environment or risk unreasonable hazard to an endangered species.

In Vitro. Literally means 'in glass', signifying a process or reaction occurring in an artificial environment, as in a test tube or culture medium.

In Vivo. Literally means 'in a living being', signifying a process or reaction occurring in a living thing.

Incentive-based Regulation. A regulation that uses the economic behavior of firms and households to attain desired environmental goals. Incentive-based programs involve taxes on emissions or tradable emission permits. The primary strength of incentive-based regulation is the flexibility it provides the polluter to find the least costly way to reduce emissions.

Incineration. A treatment technology involving destruction of waste by controlled burning at high temperatures; *e.g.*, burning sludge to remove the water and reduce the remaining residues to a safe, non-burnable ash that can be disposed of safely on land, in some waters, or in underground locations.

Incineration at Sea. Disposal of waste by burning at sea on specially-designed incinerator ships.

Incompatible Waste. A waste unsuitable for mixing with another waste or material because it may react to form a hazard.

Indirect Deforestation. Death of trees from pollution whose source is outside the forest.

Industrial Ecology. Process of designing industrial systems to behave more like ecosystems where waste from one part of the system is a resource for another part.

Industrial Melanism. A form of natural selection where the colour of a living thing helps it to blend in with an urban, industrial environment.

Industrial Sludge. Semi-liquid residue or slurry remaining from treatment of industrial water and waste-water.

Industrial Waste. Unwanted materials from an industrial operation; may be liquid, sludge, solid, or hazardous waste.

Infection. The entry, establishment and multiplication of pathogenic organisms within a host.

Infiltration. The seepage of groundwater into a sewer system, including service connections. Seepage frequently occurs through defective or cracked pipes, pipe joints, connections or manhole walls.

Influent. Water, waste-water, or other liquid flowing into a reservoir of a treatment plant.

Influent Stream. Type of stream that is everywhere above the groundwater table and flows in direct response to precipitation. Water from the channel moves down to the water table, forming a recharge mound.

Infrared Radiation. The heat energy that is emitted from all solids, liquids, and gases. In the context of the greenhouse issue, the term refers to the heat energy emitted by the Earth's surface and its atmosphere. Greenhouse gases strongly absorb this radiation in the Earth's atmosphere, and radiate some back towards the surface, creating the greenhouse effect.

Inorganic Compound. Combination of two or more elements other than those used to form organic compounds.

Inorganic Fertilizer. See *synthetic fertilizer.*

Inorganic Industrial Waste-water. Waste-water that has inorganic ions or compounds as objectionable constituents.

Inorganic Waste. Waste material such as sand, salt, iron, calcium, and other mineral materials which are only slightly affected by the action of organisms. Inorganic wastes are chemical substances of mineral origin; whereas organic wastes are chemical substances usually of animal or plant origin.

Insolation. The solar radiation incident on a unit horizontal surface at the top of the atmosphere. It is sometimes referred to as solar irradiance. The latitudinal variation of insolation supplies the energy for the general circulation of the atmosphere. Insolation depends on the angle of incidence of the solar beam and on the *solar constant.*

Instantaneous Transpiration Efficiency. A measure at the physiologic level of how well plants use available water in *photosynthesis*. The assimilation rate is divided by the transpiration rate; the moles of CO_2 taken up are divided by the moles of water lost through *transpiration* in a unit of time.

In-stream Use. A type of water use that includes navigation, generation of hydroelectric power, fish and wildlife habitat and recreation.

Integrated Energy Management. Use of a range of energy options that vary from region to region, including a mix of technology and sources of energy.

Integrated Pest Management (IPM). A mixture of chemical and other, non-pesticide, methods to control pests.

Integrated Waste Management (IWM). Set of management alternatives including reuse, source reduction, recycling, composting, landfill and incineration.

Interference. During succession, one species prevents the entrance of later successional species into an ecosystem. For example, some grasses produce such dense and thick mats that seeds of trees cannot reach the soil to germinate. As long as these grasses persist, the trees that characterise later stages of succession cannot enter the ecosystem.

Intergenerational Equity. The fairness of the distribution of the costs and benefits of a policy when costs and benefits are borne by different generations. In the case of a climate change policy the impacts of inaction in the present will be felt in future generations.

Intergovernmental Panel on Climate Change (IPCC). The IPCC was established jointly by the United Nations Environment Programme and the World Meteorological Organization in 1988. The purpose of the IPCC is to assess information in the scientific and technical literature related to all significant components of the issue of climate change. The IPCC draws upon hundreds of the world's expert scientists as authors and thousands as expert reviewers. Leading experts on climate change and environmental, social, and economic sciences from some 60 nations have helped the IPCC to prepare periodic assessments of the scientific underpinnings for understanding global climate change and its consequences. With its capacity for reporting on climate change, its consequences, and the viability of adaptation and mitigation measures, the IPCC is also looked to as the official advisory body to the world's governments on the state of the science of the climate change issue. For example, the IPCC organized the development of internationally accepted methods for conducting national greenhouse gas emission inventories.

Inversion. An anomaly in the normal positive *lapse rate;* usually refers to a thermal inversion, in which temperature increases rather than decreases with height.

Irradiance. The total radiant flux received on a unit area of a given real or imaginary surface. Also called the *radiant flux density.*

Irreversibilities. Changes that, once set in motion, cannot be reversed, at least on human time scales.

Island Arc. A curved group of volcanic islands associated with a deep-oceanic trench and subduction zone (convergent plate boundary).

Isobars. Lines drawn on a weather map joining places of equal barometric pressure.

Isopynic. A line on a chart that connects all points of equal or constant density.

Isostatic Adjustment (isostatic compensation). The process whereby lateral transport at the Earth's surface from erosion or deposition is compensated for by movements in a subcrustal layer to maintain equilibrium among units of varying masses and densities.

Isotherm. A line on a chart that connects all points of equal or constant temperature.

Isotope. One of two or more atoms that have the same atomic number (*i.e.* the same number of protons in their nuclei) but have different mass numbers.

Isthmus. Narrow strip of land located between two bodies of water; connecting two larger land areas.

J

Jet Fuel. Includes both naphtha-type and kerosene-type fuels meeting standards for use in aircraft turbine engines. Although most jet fuel is used in aircraft, some is used for other purposes such as generating electricity.

Jet Stream. Rivers of high-speed air in the atmosphere. Jet streams form along the boundaries of global air masses where there is a significant difference in atmospheric temperature. The jet streams may be several hundred miles across and 1-2 miles deep at an altitude of 8-12 miles. They usually move west to east, and are strongest in the winter with core wind speed as high as 250 mph. Changes in the jet stream indicate changes in the motion of the atmosphere and weather.

Joint Implementation. Agreements made between two or more nations under the auspices of the *United Nations Framework Convention on Climate Change* to help reduce greenhouse gas emissions.

Joule. The energy required to push with a force of one Newton for one meter.

K

Kerogen. Solid, waxy mixture of rock is heated to high temperatures, the kerogen is vaporized. The vapor is condensed and then sent to a refinery to produce gasoline, heating oil, and other products.

Kerosene. A petroleum distillate that has a maximum distillation temperature of 401 degrees Fahrenheit at the 10 per cent recovery point, a final boiling point of 572 degrees Fahrenheit, and a minimum flash point of 100 degrees Fahrenheit. Used in space heaters, cookstoves, and water heaters, and suitable for use as an illuminant when burned in wick lamps.

Keystone Species. Loosely speaking, a species, such as the sea otter, that has a large effect on its community or ecosystem so that its removal or addition to the community leads to major changes in the abundances of many or all other species.

Krakatoa (Krakatau). Active volcano 2667 ft (813 m) high in West Indonesia. It forms an island between Java and Sumatra. Its eruption in 1883, which was one of the most violent in modern times, scattered debris and darkened skies over vast areas. Additional eruptions occurred in late 1927 and in the 1960s.

Kyoto Mechanisms. The Kyoto Protocol creates three market-based mechanisms that have the potential to help countries reduce the cost of meeting their emissions reduction targets. These mechanisms are Joint Implementation (Article 6), the Clean Development Mechanisms (Article 12), and Emissions Trading (Article 17).

Kyoto Protocol. This is an international agreement struck by 159 nations attending the Third Conference of Parties

(COP) to the United Nations Framework Convention on Climate Change (held in December of 1997 in Kyoto Japan) to reduce worldwide emissions of greenhouse gases. If ratified and put into force, individual countries have committed to reduce their greenhouse gas emissions by a specified amount.

L

Lake. A body of fresh or salt water entirely surrounded by land.

Land Application. Method of disposal of hazardous waste that involves intentional application of waste material to surface soil. Useful for certain biodegradable industrial waste, such as oil and petroleum waste and some organic chemical waste.

Land Breeze. A nocturnal coastal breeze that blows from land to sea. In the evening the water may be warmer than the land, causing pressure differences. The land breeze is the flow of air from land to sea equalizing these pressure differences.

Land Ethic. A set of ethical principles that affirm the right of all resources, including plants, animals and earth materials, to continued existence and, at least in some locations, to continued existence in a natural state.

Land Use, Land-Use Change and Forestry (LULUCF). Land uses and land-use changes can act either as sinks or as emission sources. It is estimated that approximately one-fifth of global emissions result from LULUCF activities. The Kyoto Protocol allows Parties to receive emissions credit for certain LULUCF activities that reduce net emissions.

Landfill. Land waste disposal site in which waste is generally spread in thin layers, compacted, and covered with a fresh layer of soil each day.

Lapse Rate. The rapidity with which temperature decreases with altitude. The normal lapse rate is defined to be 3.6 degrees F per 1000 feet change in altitude. The dry adiabatic lapse rate is about 5.5 degrees F per 1000 feet, and the wet adiabatic lapse rate varies between 2 and 5 degrees F per 1000 feet.

Latent Heat. Energy transferred from the earth's surface to the atmosphere through the evaporation and condensation processes.

Leachate. Noxious, mineralised liquid capable of transporting bacterial pollutants. Produced when water infiltrates through waste material and becomes contaminated and polluted.

Life Zone. A climatically-defined class that can be associated with regions of soil and biota with a high uniformity in species composition and environmental adaptation.

Lifetime (Atmospheric). The lifetime of a greenhouse gas refers to the approximate amount of time it would take for the anthropogenic increment to an atmospheric pollutant concentration to return to its natural level (assuming emissions cease) as a result of either being converted to another chemical compound or being taken out of the atmosphere via a sink. This time depends on the pollutant's sources and sinks as well as its reactivity. The lifetime of a pollutant is often considered in conjunction with the mixing of pollutants in the atmosphere; a long lifetime will allow the pollutant to mix throughout the atmosphere. Average lifetimes can vary from about a week (sulfate aerosols) to more than a century (CFCs, carbon dioxide).

Light-duty Vehicles. Automobiles and light trucks combined.

Lignite. A brownish-black coal of low rank with high inherent moisture and volatile matter content, used almost exclusively for electric power generation. Also referred to as brown coal.

Limiting Factor. The single requirement for growth available in the least supply in comparison to the need of an organism. Originally applied to crops but now often applied to any species.

Liquefaction. The conversion of large solid particles of sludge into very fine particles which either dissolve or remain suspended in waste-water.

Liquefied Natural Gas (LNG). Natural gas converted to liquid form by cooling to a very low temperature.

Liquefied Petroleum Gas (LPG). Ethane, ethylene, propane, propylene, normal butane, butylene, and isobutane produced at refineries or natural gas processing plants, including plants that fractionate new natural gas plant liquids.

Lithosphere. The component of the Earth's surface comprising the rock, soil, and sediments. It is a relatively passive component of the climate system, and its physical characteristics are treated as fixed elements in the determination of climate.

Litter. Undecomposed plant residues on the soil surface.

Little Ice Age. A cold period that lasted from about A.D. 1550 to about A.D. 1850 in Europe, North America, and Asia. This period was marked by rapid expansion of mountain glaciers, especially in the Alps, Norway, Ireland, and Alaska. There were three maxima, beginning about 1650, about 1770, and 1850, each separated by slight warming intervals.

Littoral Drift. Movement caused by wave motion in nearshore and beach environment.

Littoral Zone. • That portion of a body of fresh water extending from the shoreline lakeward to the limit of occupancy of rooted plants. • A strip of land along the shoreline between the high and low water levels.

Local Extinction. The disappearance of a species from part of its range, but continued persistence elsewhere.

Loess. A buff-colored, wind-blown deposit of fine silt, which is frequently exposed in bluffs with steep faces. The thickness can range from 6 to 30 m. The loess of the USA and Europe is thought to be the fine materials first transported and deposited by the waters of melting ice sheets during the glacial period. It was later blown considerable distances with, in some cases, deposition in lakes. The origin of Asiatic loess, however, is apparently wind-blown dust from central Asian deserts.

Longwave Radiation. The radiation emitted in the spectral wavelength greater than 4 micrometers corresponding to the radiation emitted from the Earth and atmosphere. It is sometimes referred to as terrestrial radiation or infrared radiation, although somewhat imprecisely.

Low Emission Vehicle (LEV). A vehicle meeting the low-emission vehicle standards.

Low-level Nuclear Waste. Waste materials that contain sufficiently low concentrations or quantities of radioactivity so as not to present a significant environment hazard if properly handled.

Lower Heating Value. Quantity of heat liberated by the complete combustion of a unit volume or weight of a fuel assuming that the produced water remains as a vapor and the heat of the vapor is not recovered; also known as net calorific value.

Lubricant. A substance used to reduce friction between bearing surfaces or as a process material, either incorporated into other materials used as aids in manufacturing processes or as carriers of other materials. Petroleum lubricants may be produced either from distillates or residues. Other substances may be added to impart or improve useful properties. Does not include by-products of lubricating oil from solvent extraction or tars derived from de-asphalting. Lubricants include all grades of lubricating oils from spindle oil to cylinder oil and those used in greases. Lubricant categories are paraffinic and naphthenic.

M

Macronutrients. Elements required in large amounts by living things. These include the big six-carbon, hydrogen, oxygen, nitrogen, phosphorus and sulphur.

Manure. Dung and urine of animals that can be used as a form of organic fertilizer.

Marginal Land. An area of the Earth with minimal rainfall or otherwise limited severely by some necessary factor, so that it is a poor place for agriculture and easily degraded by agriculture. Typically, these lands are easily converted to deserts when used for light grazing and crop production.

Mariculture. Production of food from marine habitats.

Marine Evaporites. With respect to mineral resources, refers to materials such as potassium and sodium salts resulting from the evaporation of marine water.

Market Benefits. Benefits of a climate policy that can be measured in terms of avoided market impacts such as changes in resource productivity (*e.g.*, lower agricultural yields, scarcer water resources) and damages to human-built environment (*e.g.*, coastal flooding due to sea-level rise).

Marsh. A type of *wetland* that does not accumulate appreciable peat deposits and is dominated by herbacious vegetation. Marshes may be fresh- or saltwater, tidal or nontidal.

Mass Balance. The application of the principle of the conservation of matter. For example, the mass of a *glacier* is not destroyed or created; the mass of a glacier and all its constitutive components remains the same despite alterations in their physical states. The mass balance of a glacier is calculated with the input/output relationships of ice, firn, and snow, usually measured in water equivalent. Output includes all ablative processes of surface melting, basal melting, evaporation, wind deflation, calving, and internal melting. Input includes direct precipitation, avalanching, and the growth of superimposed ice.

Mauna Loa. An intermittently active volcano 13,680 ft (4,170 m) high in Hawaii. Last eruption was in 1984.

Mauna Loa Record. The record of measurements of atmospheric carbon dioxide concentrations taken at the Mauna Loa Observatory, Mauna Loa, Hawaii, since March 1958. The Mauna Loa record is the longest reliable daily record of atmospheric carbon dioxide measurements in the world.

Maunder Minimum. The period from 1654 to 1714 when it was believed that there were no sunspots. It is now thought that there were some sunspots during that time but less than the numbers counted after 1800.

Maximum Lifetime. Genetically determined maximum possible age to which an individual of a species can live.

Mean Sea Level. The average height of the sea surface, based upon hourly observation of the tide height on the open coast or in adjacent waters that have free access to the sea. In the United States, it is defined as the average height of the sea surface for all stages of the tide over a nineteen year period. Mean sea level, commonly abbreviated as MSL and referred to simply as sea level, serves as the reference surface for all altitudes in upper atmospheric studies.

Mesic Environment. A habitat with a moderate amount of water.

Mesoscale Eddies (mode eddies). In the ocean, dense and irregularly-oval high- and low-pressure centers about 400 km in diameter. The intensities of currents in these centers are about 10 times greater than the local means.

Mesotrophic. Reservoirs and lakes which contain moderate quantities of nutrients and are moderately productive in terms of aquatic animal and plant life.

Methane (CH_4). A hydrocarbon that is a greenhouse gas with a global warming potential most recently estimated at 21.

Methane is produced through anaerobic (without oxygen) decomposition of waste in landfills, animal digestion, decomposition of animal wastes, production and distribution of natural gas and petroleum, coal production, and incomplete fossil fuel combustion. The atmospheric concentration of methane as been shown to be increasing at a rate of about 0.6 per cent per year and the concentration of about 1.7 per million by volume (ppmv) is more than twice its pre-industrial value. However, the rate of increase of methane in the atmosphere may be stabilizing.

Methanol (CH_3OH). A colorless poisonous liquid with essentially no odor and little taste. It is the simplest alcohol with a boiling point of 64.7 degrees Celsius. In transportation, methanol is used as a vehicle fuel by itself (M100), or blended with gasoline (M85).

Methanotrophic. Having the biological capacity to oxidize methane to CO_2 and water by metabolism under aerobic conditions.

Methyl Bromide (CH_3Br). An effective pesticide; used to fumigate soil and many agricultural products. Because it contains bromine, it depletes stratospheric ozone when released to the atmosphere.

Meteorology. The science of weather-related phenomena.

Metric Ton. Common international measurement for the quantity of greenhouse gas emissions. A metric ton is equal to 2205 lbs or 1.1 short tons.

Microbial Growth. The amplification or multiplication of microorganisms such as bacteria, algae, diatoms, plankton, and fungi.

Microclimate. • Localized climate conditions within an urban area or neighborhood. • The climate around a tree or shrub or a stand of trees.

Micronutrients. Chemical elements required in very small amounts by at least some forms of life. Boron, copper and molybdenum are examples of micronutrients.

Microwave Sounding Units (MSU). Sensors carried aboard Earth orbiting satellites that have been used since 1979 to monitor tropospheric temperatures.

Migration. The movement of an individual, population or species from one habitat to another or more simply from one geographic area to another.

Milankovitch Theory. An astronomical theory formulated by the Yugoslav mathematician Milutin Milankovitch that associates climate change with fluctuations in the seasonal and geographic distribution of insolation determined by periodic variations of the Earth's eccentricity and obliquity and the longitude of the perihelion.

Mineral. Any naturally occurring inorganic substance found in the earth's crust as a crystalline solid.

Mist. Liquid particles 40 to 500 microns in diameter that are formed by condensation of *vapor* in air.

Mitigation. Process that identifies actions to avoid, lessen, or compensate for anticipated adverse environmental impacts.

Mobile Sources. Sources of air pollutants that move from place to place, for example, automobiles, trucks, buses and trains.

Model Year. Refers to the "sales" model year; for example, vehicles sold during the period from October 1 to the next September 31 is considered one model year.

Modeling. An investigative technique that uses a mathematical or physical representation of a system or theory that accounts for all or some of its known properties. Models are often used to test the effects of changes of system components on the overall performance of the system.

Molecule. Chemical combination of two or more atoms of the same chemical element (such as O_2) or different chemical elements (such as H_2O).

Monitoring. Process of collecting data on a regular basis at specific sites to provide a database from which to evaluate change. For example, collection of water samples from beneath a landfill to provide early warning should a pollution problem arise.

Monoculture. The planting of large areas with a single species or even a single strain or subspecies in farming.

Monomictic. Lakes and reservoirs which are relatively deep, do not freeze over during winter, and undergo a single stratification and mixing cycle during the year (usually in the fall).

Monsoon. A name for seasonal winds, first applied to the winds over the Arabian Sea that blow for six months from the northeast and for six months from the southwest. The term has been extended to similar winds in other parts of the world (*i.e.*, the prevailing west to northwest winds of summer in Europe have been called the European monsoon). The primary cause for these seasonal winds is the much greater annual variation of temperature over large land areas compared with neighboring ocean surfaces, causing an excess of pressure over the continents in winter and a deficit in summer, but other factors, such as topography of the land, also have an effect. The monsoons are strongest in the southern and eastern sides of Asia, but also occur along the coasts of tropical regions wherever the planetary circulation is not strong enough to inhibit them. The monsoon climate can be described as a long winter-spring dry season, which includes a cold season followed by a short hot season just preceding the rains; a summer and early autumn rainy season, which is generally very wet but may vary greatly from year to year; and a secondary warming immediately after the rainy season.

Montreal Protocol on Substances that Deplete the Ozone Layer. The Montreal Protocol and its amendments control the phaseout of ozone depleting substances production and use. Under the Protocol, several international organizations report on the science of ozone depletion, implement projects to help move away from ozone depleting substances, and provide a forum for policy discussions. In the United States, the Protocol is implemented under the Clean Air Act Amendments of 1990.

Motor Gasoline. A complex mixture of relatively volatile hydrocarbons, with or without small quantities of additives, obtained by blending appropriate refinery streams to form a fuel suitable for use in spark-ignition engines. Motor gasoline includes both leaded and unleaded grades of finished gasoline, blending components, and gasohol.

Mount Pinatubo. A volcano in the Philippine Islands that erupted in 1991. The eruption of Mount Pinatubo ejected enough particulate and sulfate aerosol matter into the atmosphere to block some of the incoming solar radiation from reaching Earth's atmosphere. This effectively cooled the planet from 1992 to 1994, masking the warming that had been occurring for most of the 1980s and 1990s.

Muck Soils. Earth made from decaying plant materials.

Mulch. A layer of material (wood chips, straw, leaves, etc.) placed around plants to hold moisture, prevent weed growth, and enrich or sterilize the soil.

Municipal Solid Waste (MSW). Residential solid waste and some non-hazardous commercial, institutional, and industrial wastes. This material is generally sent to municipal landfills for disposal.

N

Naphtha. A generic term applied to a petroleum fraction with an approximate boiling range between 122 and 400 degrees Fahrenheit.

National Action Plans. Plans submitted to the Conference of the Parties (COP) by all Parties outlining the steps that they have adopted to limit their anthropogenic GHG emissions. Countries must submit these plans as a condition of participating in the UN Framework Convention on Climate Change and, subsequently, must communicate their progress to the COP regularly.

Natural Catastrophe. Sudden change in the environment, not the result of human actions.

Natural Gas. Underground deposits of gases consisting of 50 to 90 per cent methane (CH_4) and small amounts of heavier gaseous hydrocarbon compounds such as propane (C_3H_4) and butane (C_4H_{10}).

Natural Gas Liquids (NGLs). Those hydrocarbons in natural gas that are separated as liquids from the gas. Includes natural gas plant liquids and lease condensate.

Natural Selection. The process of survival of the fittest by which organisms that adapt to their environment survive while those that do not adapt disappear.

Nature Preserve. An area set aside with the primary purpose of conserving some biological resource.

Necrosis. Death of plant or animal cells or tissues. In plants, necrosis can discolor stems or leaves or kill a plant entirely.

Negative Declaration. Document that may be filed if an agency has determined that a particular project will not have significant adverse effect on the environment.

Negative Feedback. An interaction that reduces or dampens the response of the system in which it is incorporated.

Nepheloccygia. Clouds that resemble recognizable shapes.

Net Primary Production. The part of the gross primary production that remains stored in the producer organism (primarily green plants) after deducting the amount used during the process of respiration. Abbreviated NPP.

New Forestry. The name for a new variety of timber harvesting practices to increase the likelihood of sustainability, including recognition of the dynamic characteristics of forests and of the need for management within an ecosystem context.

Niche. (i) The 'profession' or role, of an organism or species; and (ii) all the environmental conditions under which the individual or species can persist. The *fundamental niche* is all the conditions under which a species can persist in the absence of competition; the *realised niche* is the set of conditions as they occur in the real world with competitors.

Nitrate. A compound containing nitrogen that can exist in the atmosphere or as a dissolved gas in water and which can have harmful effects on humans and animals. Nitrates in water can cause severe illness in infants and domestic animals. A plant nutrient and inorganic fertilizer, nitrate is found in septic systems, animal feed lots, agricultural fertilizers, manure, industrial waste-waters, sanitary landfills, and garbage dumps.

Nitrogen Cycle. Cyclic movement of nitrogen in different chemical forms from the environment, to organisms, and then back to the environment.

Nitrogen Fixation. Conversion of atmospheric nitrogen gas into forms useful to plants and other organisms by lightning, bacteria, and blue-green algae; it is part of the nitrogen cycle.

Nitrogen Oxides (NO_x). Gases consisting of one molecule of nitrogen and varying numbers of oxygen molecules. Nitrogen oxides are produced, for example, by the combustion of fossil fuels in vehicles and electric power plants. In the atmosphere, nitrogen oxides can contribute to formation of photochemical ozone (smog), impair visibility, and have health consequences; they are considered pollutants.

Nitrous Oxide (N_2O). A powerful greenhouse gas with a global warming potential most recently evaluated at 310. Major sources of nitrous oxide include soil cultivation practices, especially the use of commercial and organic fertilizers, fossil fuel combustion, nitric acid production, and biomass burning.

Noise Pollution. A type of pollution characterised by unwanted or potentially damaging sound.

Non-biodegradable. Substance that cannot be broken down in the environment by natural processess.

Non-linearities. Occur when changes in one variable cause a more than proportionate impact on another variable.

Non-renewable Resource. A resource that is cycled so slowly by natural Earth processes that once used, it is essentially not going to be made available within any useful time framework.

Non-Annex I Parties. Countries that have ratified or acceded to the UNFCCC that are listed in Annex I of the UNFCCC.

Non-Annex B Parties. Countries that are not listed in Annex B of the Kyoto Protocol.

Non-market Benefits. Benefits of a climate policy that can be measured in terms of avoided non-market impacts such as human-health impacts (*e.g.*, increased incidence of tropical diseases) and damages to ecosystems (*e.g.*, loss of biodiversity).

Non-methane Volatile Organic Compounds (NMVOCs). Organic compounds, other than methane, that participate in atmospheric photochemical reactions.

Non-party. A state that has not ratified the UNFCCC. Non-parties may attend talks as observers.

Non-point Sources. Diffuse pollution sources (*i.e.* without a single point of origin or not introduced into a receiving stream from a specific outlet). The pollutants are generally carried off the land by storm water. Common non-point sources are agriculture, forestry, urban, mining, construction, dams, channels, land disposal, saltwater intrusion, and city streets.

Non-potable. Water that is unsafe or unpalatable to drink because it contains pollutants, contaminants, minerals, or infective agents.

Non-renewable Energy. Alternative energy sources, including nuclear and geothermal, that are dependent on fuels or a resource that may be used up much faster than it is replenished by natural processes.

Non-road Emissions. Pollutants emitted by combustion engines on farm and construction equipment, gasoline-powered lawn and garden equipment, and power boats and outboard motors.

Nuclear Electric Power. Electricity generated by an electric power plant whose turbines are driven by steam generated in a reactor by heat from the fissioning of nuclear fuel.

Nuclear Energy. Energy released when atomic nuclei undergo a nuclear reaction such as the spontaneous emission of radioactivity, nuclear fission, or nuclear fusion. Controlled nuclear fission reactions take place within commercial nuclear reactors to produce energy.

Nuclear Winter. Prediction by some scientists that smoke and debris rising from massive fires of a nuclear war could block sunlight for weeks or months, cooling the earth's surface and producing climate changes that could, for example, negatively affect world agricultural and weather patterns.

Nutrient. Any substance assimilated by living things that promotes growth.

Nutrient Pollution. Contamination of water resources by excessive inputs of nutrients. In surface waters, excess algal production is a major concern.

O

Oasis. A spot in a desert made fertile by water; which normally originates as groundwater.

Occluded Front (occlusion). A composite of two fronts formed as a cold front overtakes a warm front. A cold occlusion results when the coldest air is behind the cold front. The cold front undercuts the warm front and, at the Earth's surface, coldest air replaces less-cold air.

A warm occlusion occurs when the coldest air lies ahead of the warm front. Because the cold front cannot lift the colder air mass, it rides piggyback up on the warm front over the coldest air

Ocean. The salt water surrounding the great land masses. The land masses divide the ocean into several distinct portions, each of which also is called an ocean. The oceans include the Pacific Ocean, the Atlantic Ocean, the Indian Ocean, and the Arctic Ocean.

Ocean Mixing. Processes that involve rates of advection, upwelling/downwelling, and eddy diffusion and that determine how rapidly excess atmospheric carbon dioxide can be taken up by the oceans.

Ocean Thermal Conversion. Direct utilisation of solar energy using part of a natural oceanic environment as a gigantic solar collector.

Off-Site Effect. An environmental effect occurring away from the location of the causal factors.

Off-Site Facility. A hazardous waste treatment, storage or disposal area that is located away from the generating site.

Offsets. A concept whereby emissions from proposed new or modified stationary sources are balanced by reductions from existing sources to stabilize total emissions.

Offstream Use. Water withdrawn from surface or groundwater sources for use at another place.

Oil and Gas Waste. Gas and oil drilling muds, oil production brines, and other waste associated with exploration for, development and production of crude oil or natural gas.

Oil shale. Underground formation of a fine-grained sedimentary rock containing varying amounts of kerogen, a solid, waxy mixture of hydrocarbon compounds. Heating the rock to high temperatures converts the kerogen to a vapor, which can be condensed to form a slow flowing heavy oil called shale oil.

Oil Spill. An accidental or intentional discharge of oil which reaches bodies of water. Can be controlled by chemical dispersion, combustion, mechanical containment, and/or adsorption. Spills from tanks and pipelines can also occur away from water bodies, contaminating the soil, getting into sewer systems and threatening underground water sources.

Old Growth. A nontechnical term often used to mean a virgin forest (one never cut) but also used to mean a forest that has been undisturbed for a long, but usually unspecified, time.

Oligotrophic Lakes. Deep clear lakes with few nutrients, little organic matter and a high dissolved-oxygen level.

Omnivores. Organisms that eat both plants and animals.

On-Site Effect. An environmental effect occurring at the location of the causal factors.

On-Site Facility. A hazardous waste treatment, storage or disposal area that is located on the generating site.

Opacity. The degree of obscuration of light; for example, a glass window has almost 0% opacity, whereas a concrete wall has 100% opacity.

Open Dumps. Area where solid waste is disposed of by simply dumping it. Often causes severe environmental problems, such as water pollution and creates a health hazard.

Open Woodlands. Areas in which trees are a dominant vegetation form but the leaves of adjacent trees generally do not touch or overlap, so that there are gaps in the canopy. Typically, grasses or shrubs grow in these gaps among the trees.

Optical Thickness (optical depth). In calculating the transfer of radiant energy, the mass of an absorbing or emitting material lying in a vertical column of unit cross-sectional area and extending between two specified levels. Also, the degree to which a cloud prevents light from passing through it; the optical thickness then depends on the physical constitution (crystals, drops, and/or droplets), the form, the concentration, and the vertical extent of the cloud.

Optimal Carrying Capacity. This term has several meanings, but the major idea is the maximum abundance of a population or species that can persist in an ecosystem without degrading the ability of the ecosystem to maintain: (i) that population or species; (ii) all necessary ecosystem processes; and (iii) the other species found in that ecosystem.

Optimum Sustainable Population (OSP). The population level that results in an optimum sustainable yield; or the population level that is in some way best for the population, its ecological community, its ecosystem, or the biosphere.

Optimum Sustainable Yield (OSY). The largest yield of a renewable resource achievable over a long time period without decreasing the ability of the population or its environment to support the continuation of this level of yield.

Ore. Mineral deposit containing a high enough concentration of at least one metallic element to permit the metal to be extracted and sold at a profit.

Organic. Referring to or derived from living organisms.

Organic Compound. Molecule that contains atoms of the element carbon, usually combined with itself and with atoms of one or more other element such as hydrogen, oxygen, nitrogen, sulfur, phosphorus, chlorine, or fluorine.

Organic Fertilizer. Organic material such as manure or compost, applied to cropland as a source of plant nutrients.

Organic Industrial Waste-water. Industrial waste-water that has organic compounds as the objectionable constituents.

Organic Matter. Carbonaceous waste contained in plant or animal matter and originating from domestic or industrial sources.

Organic Waste. Waste material which comes-mainly from animal or plant sources. Organic waste generally can be consumed by bacteria and other small organisms. Inorganic wastes are chemical substances of mineral origin.

Organism. Any form of animal or plant life.

Osmosis. The diffusion of water across a selectively permeable membrane in the direction of lower water concentration.

Overdraft. Groundwater withdrawal when the amount pumped from wells exceeds the natural rate of replenishment.

Overgrazing. When the carrying capacity of land for an herbivore, such as cattle or deer, is exceeded.

Oxidants. Those substances in the air (such as ozone, PAN and nitrogen dioxide) that are capable of oxidising other chemicals or elements in oxidation-reduction type chemical reactions. Oxidants are made up mostly of ozone and only small amounts of the other materials.

Oxidize. To chemically transform a substance by combining it with oxygen.

Oxygen Cycle. Cyclic movement of oxygen in different chemical forms from the environment, to organisms, and then back to the environment.

Ozone (O_3). A colorless gas with a pungent odor, having the molecular form of O_3, found in two layers of the atmosphere, the stratosphere (about 90% of the total atmospheric loading) and the troposphere (about 10%). Ozone is a form of oxygen found naturally in the stratosphere that provides a protective layer shielding the Earth from ultraviolet radiation's harmful health effects on humans and the environment. In the troposphere, ozone is a chemical oxidant and major component of photochemical smog. Ozone can seriously affect the human respiratory system.

Ozone Depleting Substance (ODS). A family of man-made compounds that includes, but are not limited to, chlorofluorocarbons (CFCs), bromofluorocarbons (halons), methyl chloroform, carbon tetrachloride, methyl bromide, and hydrochlorofluorocarbons (HCFCs). These compounds have been shown to deplete stratospheric ozone, and therefore are typically referred to as ODSs.

Ozone Hole. A large area of intense stratospheric ozone depletion over the Antarctic continent that typically occurs annually between late August and early October; and generally ends in mid-November. This severe ozone thinning has increased conspicuously since the late seventies and early eighties. This phenomenon is the result of chemical mechanisms initiated by man-made chlorofluorocarbons (see CFCs). Continued buildup of CFCs is expected to lead to additional ozone loss worldwide.

The thinning is focused in the Antarctic because of particular meteorological conditions there. During Austral spring (September and October in the Southern Hemisphere) a belt of stratospheric winds encircles Antarctica essentially isolating the cold stratospheric air there from the warmer air of the middle latitudes. The frigid air permits the formation of ice clouds that facilitate chemical interactions among nitrogen, hydrogen, and chlorine (elevated from CFCs) atoms, the end product of which is the destruction of ozone.

Ozone Layer. The layer of gaseous ozone (O_3) in the stratosphere that protects life on earth by filtering out harmful ultraviolet radiation from the sun.

Ozone Precursors. Chemical compounds, such as carbon monoxide, methane, non-methane hydrocarbons, and nitrogen oxides, which in the presence of solar radiation react with other chemical compounds to form ozone, mainly in the troposphere.

Ozone Shield. Stratospheric ozone layer that absorbs ultraviolet radiation.

P

Paleoclimate. Climate as it existed in the distant past, particularly before historical records.

Paleosol. An ancient soil or soil horizon that formed on the surface during the geologic past.

Paludification. The expansion of a bog caused by the gradual rising of the water table as accumulation of peat impedes water drainage.

Palynology. The science of reconstructing the past flora and past climate from pollen data obtained from lake and bog sediments. The fossil pollen record is a function of the regional flora and vegetation at a given time and location.

Particulates. 1. Fine liquid or solid particles such as dust, smoke, mist, fumes, or smog, found in air or emissions. 2. Very small solids suspended in water; they can vary in size, shape, density and electrical charge and can be gathered together by coagulation and flocculation.

Parts per billion (ppb). Number of parts of a chemical found in one billion parts of a particular gas, liquid, or solid mixture.

Parts per million (ppm). Number of parts of a chemical found in one million parts of a particular gas, liquid, or solid.

Past Climate Analogs. The reconstructing of past climates at a given locality from modern climatic conditions in a different elevation or latitudinal zone to infer past climatic conditions.

Pasture. Land plowed and planted to provide forage for domestic herbivorous animals.

Pathogens. Microorganisms (*e.g.*, bacteria, viruses, or parasites) that can cause disease in humans, animals and plants.

pCO_2. The partial pressure of CO_2 in the atmosphere and the ocean. In the atmosphere, the partial pressure of CO_2 is defined as the pressure the CO_2 would exert if all other gases were removed. The sum of the partial pressure of all the atmospheric gases will equal the atmospheric pressure. The partial pressure of CO_2 in the atmosphere is determined by the atmospheric CO_2 concentration and atmospheric temperature. In the ocean, the pCO_2 is determined by the amount of dissolved CO_2 and H_2CO_3. It varies with alkalinity, latitude, depth, and temperature. Biological processes in the ocean also exert an influence on the pCO_2 in the ocean.

Peak Levels. Levels of airborne pollutant contaminants much higher than average or occurring for short periods of time in response to sudden releases.

Pedology. The study of soils.

Pentanes Plus. A mixture of hydrocarbons, mostly pentanes and heavier fractions, extracted from natural gas.

Perched Water. Zone of unpressurized water held above the water table by impermeable rock or sediment.

Percolating Water. Water that passes through rocks or soil under the force of gravity.

Percolation. The movement of water downward and radially through the subsurface soil layers, usually continuing downward to the *groundwater.*

Perfluorocarbons (PFCs). A group of human-made chemicals composed of carbon and fluorine only. These chemicals (predominantly CF_4 and C_2F_6) were introduced as alternatives, along with hydrofluorocarbons, to the ozone depleting substances. In addition, PFCs are emitted as by-products of industrial processes and are also used in manufacturing. PFCs do not harm the stratospheric ozone layer, but they are powerful greenhouse gases. CF_4 has a global warming potential (GWP) of 6,500 and C_2F_6 has a GWP of 9,200.

Permafrost. Perennially frozen ground that occurs wherever the temperature remains below 0 degrees C for several years.

Pest. An insect, rodent, nematode, fungus, weed or other form of terrestrial or aquatic plant or animal life that is injurious to health or the environment.

Pesticide. Substances or mixture there of intended for preventing, destroying, repelling, or mitigating any pest. Also, any substance or mixture intended for use as a plant regulator, defoliant, or desiccant.

Petrochemical Feedstock. Feedstock derived from petroleum, used principally for the manufacture of chemicals, synthetic rubber, and a variety of plastics. The categories reported are naphtha (endpoint less than 401 degrees Fahrenheit) and other oils (endpoint equal to or greater than 401 degrees Fahrenheit).

Petrochemicals. Chemicals obtained by refining (*i.e.*, distilling) crude oil. They are used as raw materials in the manufacture of most industrial chemicals, fertilizers, pesticides, plastics, synthetic fibers, paints, medicines, and many other products.

Petroleum. A generic term applied to oil and oil products in all forms, such as crude oil, lease condensate, unfinished oils, petroleum products, natural gas plant liquids, and non-hydrocarbon compounds blended into finished petroleum products.

Petroleum Coke. A residue that is the final product of the condensation process in cracking.

Phenology. The study of periodic biological phenomena with relation to climate, particularly seasonal changes. These phenomena can be used to interpret local seasons and the climatic zones.

Phosphorus Cycle. Major biogeochemical cycle involving the movement of phosphorus throughout the biosphere and lithosphere. This cycle is important because phosphorus is an essential element for life and often is a limiting nutrient for plant growth.

Photochemical Oxidants. Air pollutants formed by the action of sunlight on oxides of nitrogen and hydrocarbons.

Photochemical Reactions. Chemical reactions occuring in the atmosphere under the influence of sunlight. The energy for these reactions is obtained from sunlight, particularly ultraviolet rays. The results of these reactions are photochemical smogs.

Photochemical Smog. Air pollution caused by chemical reactions among various substances and pollutants in *the atmosphere.*

Photoelectric. Of or relating to the electrical effects of light, including the emission of electrons, the generation of a voltage, or a change in resistance.

Photosynthesis. The manufacture by plants of carbohydrates and oxygen from carbon dioxide and water in the presence of chlorophyll with sunlight as the energy source. Oxygen and water vapor are released in the process. Photosynthesis is dependent on favorable temperature and moisture conditions as well as on the atmospheric carbon dioxide concentration. Increased levels of carbon dioxide can increase net photosynthesis in many plants.

Photovoltaic and Solar Thermal Energy. Energy radiated by the sun as electromagnetic waves (electromagnetic radiation) that is converted into electricity by means of solar (*i.e.*, photovoltaic) cells or useable heat by concentrating (*i.e.*, focusing) collectors.

Physiographic Province. A region characterised by a particular assemblage of landforms, climate and geomorphic history.

Phytoplankton. That portion of the *plankton* community comprised of tiny plants (*e.g.*, *algae* and diatoms).

Phytotoxic. Harmful to plants.

Phytotreatment. The cultivation of specialized plants that absorb specific contaminants from the soil through their roots or foliage. This reduces the concentration of contaminants in the soil, but incorporates them into biomasses that may be released back into the environment when the plant dies or is harvested.

Pioneer Species. Species found in early stages of succession.

Placer Deposit. A type of ore deposit found in material transported and deposited by agents such as running water, ice or wind; for example, gold and diamonds found in stream deposits.

Planetary Albedo. The fraction (approximately 30%) of incident solar radiation that is reflected by the earth-atmosphere system and returned to space, mostly by backscatter from clouds in *the atmosphere.*

Planetary Boundary Layer. The transition region between the turbulent surface layer and the normally nonturbulent free *atmosphere.* This region is about 1 km in thickness and is characterized by a well-developed mixing generated by frictional drag as the air masses move over the Earth's surface. This layer contains approximately 10% of the mass of the atmosphere. Also called the atmospheric boundary layer or frictional layer.

Plankton. Passively floating or weakly motile aquatic plants (*phytoplankton*) and animals (**zooplankton**).

Plate Tectonics. A model of global tectonics that suggests that the outer layer of Earth, known as the *lithosphere,* is composed of several large plates that move relative to one another. Continents and oceans basins are passive riders on these plates.

Pleistocene. The earlier of the two epochs of the *Quaternary period,* starting 2 to 3 million years before the present and ending about 10,000 years ago. It was a time of glacial activity.

Point Source. A single identifiable source that discharges pollutants into the environment. Examples are smokestack, sewer, ditch, or pipe.

Polar Stratospheric Clouds. Clouds that form in the stratosphere during the polar winter.

Polar Vortex. Arctic air masses that in the winter become isolated from the rest of the atmosphere and circulate about the pole. The vortex rotates counter clockwise because of the rotation of the Earth in the Southern Hemisphere.

Pollutant. In general terms, any factor that has a harmful effect on living things or their environment.

"Polluter Pays" Principle (PPP). The principle that countries should in some way compensate others for the effects of pollution that they (or their citizens) generate or have generated.

Pollution. A change in the physical, chemical, or biological characteristics of the air, water, or soil that can affect the health, survival, or activities of humans in an unwanted way. Some expand the term to include harmful effects on all forms of life.

Polyvinyl Chloride (PVC). A polymer of vinyl chloride. It is tasteless, odorless and insoluble in most organic solvents. A member of the family vinyl resin, used in soft flexible films for food packaging and in molded rigid products, such as pipes, fibers, upholstery, and bristles.

Population. Group of individual organisms of the same species living within a defined area.

Porosity. Degree to which soil, gravel, sediment, or rock is permeated with pores or cavities through which water or air can move.

Potable. Describing water that is relatively clear, odour-free and safe to drink.

Positive Feedback. An interaction that amplifies the response of the system in which it is incorporated.

ppm. Parts per million—the unit commonly used to designate the concentration of a substance in a waste-water in terms of weight, *i.e.* one pound per million pounds, etc., ppm is synonymous with the more commonly used term mg/l (milligrams per liter).

ppm or ppb. Abbreviations for "parts per million" and "parts per billion," respectively—the units in which concentrations of greenhouse gases are commonly presented. For example, since the pre-industrial era, atmospheric concentrations of carbon dioxide have increased from 270 ppm to 370 ppm.

Precession. The tendency of the Earth's axis to wobble in space over a period of 23,000 years. The Earth's precession is one of the factors that results in the planet receiving different amounts of *solar energy* over extended periods of time.

Precipitation. Any or all forms of liquid or solid water particles that fall from the atmosphere and reach the Earth's surface. It includes drizzle, rain, snow, snow pellets, snow grains, ice crystals, ice pellets, and hail. The ratio of precipitation to evaporation is the most important factor in the distribution of vegetation zones. Precipitation is also defined as a measure of the quantity, expressed in centimeters or milliliters of liquid water depth, of the water substance that has fallen at a given location in a specified amount of time.

Predation-parasitism. Interaction between individuals of two species in which the outcome benefits one and is detrimental to the other.

Predator. An organism that feeds on other, live organisms, usually of other species. The term is usually applied to animals that feed on other animals, but sometimes it is used to mean herbivore.

Prescribed Burning. Deliberate setting and careful control of surface fires in forests to help prevent more destructive fires and to kill off unwanted plants that compete with commercial species for plant nutrients; may also be used on grasslands.

Primary Oil Recovery. Pumping out the crude oil that flows by gravity into the bottom of an oil well.

Primary Pollutants. Air pollutants emitted directly into the atmosphere. Included are particulates, sulphur oxides, carbon monoxide, nitrogen oxides and hydrocarbons.

Primary Succession. The natural development of vegetation and soil on a site that had not previously borne vegetation (*e.g.*, a sand dune or lava flow), which vegetation will be replaced by other, successive plant communities.

Primitive Equations. The Eulerian equations of fluid motion in which the primary dependent variables are the velocity components of the fluid. In meteorology, they can be specialized to apply directly to the cylonic-scale motions.

Production, Ecological. The amount of increase in organic matter, usually measured per unit area of land surface or unit volume of water, as in grams per square meter (g/m^2). Production is divided into primary (that of autotrophs) and secondary (that of heterotrophs). It is also divided into net (that which remains stored after use) and gross (that added before any use).

Production, Primary. The production by autotrophs.

Production, Secondary. The production by heterotrophs.

Productivity, Ecological. The rate of production; that is, the amount of increase in organic matter per unit time (for example, grams per meter squared per year).

Prokaryote. A kind of organism that lacks a true cell nucleus and has other cellular characteristics that distinguish it from the eukaryotes. Bacteria and blue-green algae are prokaryotes.

Protocooperation. A symbiotic relationship that is beneficial, but not obligatory, to both species.

Proxy Climate Indicators. Dateable evidence of a biological or geological phenomenon whose condition, at least in part, is attributable to climatic conditions at the time of its formation. Proxy data are any material that provides an indirect measure of climate and include documentary evidence of crop yields, harvest dates, glacier movements, tree rings, varves, glaciers and snow lines, insect remains, pollen remains, marine microfauna, isotope measurements. 18O, in ice sheets, 18O, 2H, and 13C in tree rings; $CaCO_3$ in sediments; and speleothems. There are three main problems in using proxy data. (1) dating, (2) lag and response time, and (3) meteorological interpretation. Tree rings, pollen deposits from varved lakes, and ice cores are the most promising proxy data sources for reconstructing the climate of the last five millennia because the dating are precise on an annual basis while other proxy data sources may only be precise to +/- 100 years.

Putrefaction. Biological decomposition of organic matter; associated with anaerobic conditions.

Putrescible. Able to rot quickly enough to cause odors and attract flies.

Pycnocline. In the ocean, a region where the water density increases rapidly with depth.

Pyrgeometer. An instrument that measures radiation from the earth's surface into space.

Pyrolysis. Breaking down burnable waste by combustion in the absence of oxygen. Usually, high heat is applied to the waste in a closed chamber evaporating all moisture and breaking down materials into gaseous hydrocarbons and carbon-like residue.

Q

QELRC (Quantified Emission Limitation and Reduction Commitment). Also known as QELRO (Quantified Emission Limitation and Reduction Objective). The quantified commitments for GHG emissions listed in Annex B of the Kyoto Protocol. QELRCs are specified in percentages relative to 1990 emissions.

Quaternary Period. The latest period of geologic time, covering the most recent 2,000,000 years of the Earth's history. It is divided into two epochs: the *Pleistocene*—2 million years ago to approximately 10,000 years ago—and the *Holocene*—the period from approximately 10,000 years ago to the present. The Quaternary period is the artificial division of time separating prehuman and human periods. It contains five ice ages and four interglacial ages, and temperature indicators seem to show sharp and abrupt changes by several degrees.

R

Radiant Flux Density. The total flow of radiation received on a unit area of a given real or imaginary surface. Also called the *irradiance.*

Radiation. Energy emitted in the form of electromagnetic waves. Radiation has differing characteristics depending upon the wavelength. Because the radiation from the Sun is relatively energetic, it has a short wavelength (ultraviolet, visible, and near infrared) while energy radiated from the Earth's surface and the atmosphere has a longer wavelength (*e.g.*, infrared radiation) because the Earth is cooler than the Sun.

Radiation Balance. The difference between the absorbed solar radiation and the net *infrared radiation.* Experimental data show that radiation from the earth's natural surfaces is rather close to the radiation from a black body at the corresponding temperature; the ratio of the observed values of radiation to black body radiation is generally 0.90-1.0.

Radiative-convective Models. Thermodynamic models that determine the equilibrium temperature distribution for an atmospheric column and the underlying surface, subject to prescribed solar radiation at the top of the atmosphere and prescribed atmospheric composition and surface albedo. Submodels for the transfer of solar and terrestrial radiation, the heat exchange between the earth's surface and atmosphere, the vertical redistribution of heat within the atmosphere, the atmospheric water vapor content and clouds are included in these one-dimensional models. Abbreviated as RCM.

Radiative Forcing. A change in the balance between incoming solar radiation and outgoing infrared radiation. Without any radiative forcing, solar radiation coming to the Earth would continue to be approximately equal to the infrared radiation emitted from the Earth. The addition of greenhouse gases traps an increased fraction of the infrared radiation, radiating it back toward the surface and creating a warming influence (*i.e.* positive radiative forcing because incoming solar radiation will exceed outgoing infrared radiation).

Radiatively Active Gases. Gases that absorb incoming solar radiation or outgoing *infrared radiation,* thus affecting the vertical temperature profile of the atmosphere. Most frequently being cited as being radiatively active gases are water vapor, CO_2, methane, nitrous oxide, chlorofluorocarbons, and ozone.

Radioactive Decay. A process of decay of radioisotopes that change from one isotope to another and emit one or more forms of radiation.

Radioactive Waste. Any waste that emits energy as rays, waves, streams or energetic particles. Radioactive materials are often mixed with hazardous waste, from nuclear reactors, research institutions, or hospitals.

Radioisotope. A form of a chemical element that spontaneously undergoes radioactive decay.

Radiosonde. A balloon-borne instrument for the simultaneous measurement and transmission of meteorological data up to a height of approximately 30,000 meters (100,000 feet). The height of each pressure level of the observation is computed from data received via radio signals.

Radon. Naturally occurring radioactive gas. Radon is colourless, odourless and tasteless and must be identified through proper testing.

Rangeland. Land, mostly grasslands, whose plants can provide food (*i.e.*, forage) for grazing or browsing animals.

Raw Sewage. Untreated waste-water and its contents.

Re-entry. (In indoor air program) Refers to air exhausted from a building that is immediately brought back into the system through the air intake and other openings.

Reactivity. Refers to those hazardous wastes that are normally unstable and readily undergo violent chemical change but do not explode.

Reaeration. Introduction of air into the lower layers of a reservoir. As the air bubbles form and rise through the water, the oxygen dissolves into the water and replenishes the dissolved oxygen. The rising bubbles also cause the lower waters to rise to the surface where they take on oxygen from the atmosphere.

Receiving Waters. A river, lake, ocean, stream or other watercourse into which waste-water or treated effluent is discharged.

Recharge. The process by which water is added to a *reservoir* or zone of saturation, often by *runoff* or *percolation* from the soil surface.

Recharge Area. A land area in which water reaches the zone of saturation from surface infiltration, *e.g.*, where rainwater soaks through the earth to reach an aquifer.

Recycling. Collecting and reprocessing a resource so it can be used again. An example is collecting aluminum cans, melting them down, and using the aluminum to make new cans or other aluminum products.

Reduce, Reuse and Recycle. The three R's of integrated waste management.

Reflective Environment. An environment in which large sound absorptive surfaces are present.

Reflectivity. The ratio of the energy carried by a wave that is reflected from a surface to the energy of a wave incident on the surface.

Reforestation. Replanting of forests on lands that have recently been harvested.

Relative Sea Level. The height of the boundary between sea and air as measured in relationship to a fixed reference point on land.

Renewable Energy. Energy obtained from sources that are essentially inexhaustible, unlike, for example, the fossil fuels, of which there is a finite supply. Renewable sources of energy include wood, waste, geothermal, wind, photovoltaic, and solar thermal energy.

Renewable Resource. A resource, such as timber, water, or air, that is naturally recycled or recycled by artificial processes within a time framework useful for people.

Representative Natural Areas. Parks or preserves set aside to represent presettlement conditions of a specific ecosystem type.

Reserves. Known and identified deposits of earth materials from which useful materials can be extracted profitably with existing technology and under present economic and legal conditions.

Reservoir. Any natural or artificial holding area used to store, regulate, or control a substance.

Residence Time. The average time spent in a reservoir by an individual atom or molecule. With respect to greenhouse gases, residence time usually refers to how long a particular molecule remains in the atmosphere.

Residential Sector. An area or portion consisting only of housing units.

Residential Waste. Waste generated in single and multi-family homes, including newspapers, clothing, disposable tableware, food packaging, cans, bottles, food scraps, and yard trimmings other than those that are diverted to backyard composting.

Residual. Amount of a pollutant remaining in the environment after a natural or technological process has taken place; *e.g.*, the sludge remaining after initial waste-water treatment, or particulates remaining in air after it passes through a scrubbing or other process.

Residual Fuel Oil. The heavier oils that remain after the distillate fuel oils and lighter hydrocarbons are distilled away in refinery operations and that conform to ASTM Specifications D396 and D975. Included are No. 5, a residual fuel oil of medium viscosity; Navy Special, for use in steam-powered vessels in government service and in shore power plants; and No. 6, which includes Bunker C fuel oil and is used for commercial and industrial heating, electricity generation, and to power ships. Imports of residual fuel oil include imported crude oil burned as fuel.

Residual Risk. The extent of health risk from air pollutants remaining after application of the Maximum Achievable Control Technology (MACT).

Resistance. For plants and animals, the ability to withstand poor environmental conditions or attacks by chemicals or disease. May be inborn or acquired.

Resource-based Agriculture. Agriculture with production limited by the availability of resources.

Resources. Reserves plus other deposits of useful earth materials that may eventually become available.

Respiration. A biochemical process by which living organisms take up oxygen from the environment and consume organic matter, releasing both carbon dioxide and heat. In plants, the organic matter in photosynthate produced during daylight hours.

Returned Sludge. Settled activated sludge that is returned to mix with the raw or primary settled waste-water.

Revelle Factor. The ratio of the instantaneous fractional change in the partial pressure of CO_2 (pCO_2) exerted by seawater to the fractional change in total CO_2 dissolved in the ocean waters. The buffer factor relates the partial pressure of CO_2 in the ocean to the total ocean CO_2 concentration at constant temperature, alkalinity and salinity. The Revelle factor is a useful parameter for examining the distribution of CO_2 between the atmosphere and the ocean, and measures in part the amount of CO_2 that can be dissolved in the mixed surface layer.

Revenue Recycling. If permits are auctioned, this gives considerable sums of money to be recycled back into the economy, either through a lump sum payment of offsetting other taxes. If the existing taxes that are correspondingly reduced were very inefficient, this allows the possibility of both environmental and economic benefits from the trading system, commonly called the 'double dividend.'

Reverse Osmosis. A treatment process used in water systems by adding pressure to force water through a semi-permeable membrane. Reverse osmosis removes most drinking water contaminants. Also used in waste-water treatment. Large-scale reverse osmosis plants are being developed.

Riffle. A section of stream channel characterised at low flow by fast, shallow flow. Generally contains relatively coarse bed-load particles.

Risk (Adverse) for Endangered Species. Risk to aquatic species if anticipated pesticide residue levels equal one-fifth of LD10 or one-tenth of LC50; risk to terrestrial species if anticipated pesticide residue levels equal one-fifth of LC10 or one-tenth of LC50.

Risk Assessment. Qualitative and quantitative evaluation of the risk posed to human health and/or the environment by the actual or potential presence and/or use of specific pollutants.

River Basin. The land area drained by a river and its tributaries.

Rocketsonde. A rocket-borne instrument for measurement and transmission of upper-air meteorological data in the lower 76,000 meters (250,000 feet) of the atmosphere, especially that portion inaccessible to *radiosonde* techniques.

Rule of Climatic Similarity. Similar environment lead to the evolution of organisms similar in form and function (but not necessarily in genetic heritage or internal makeup) and to similar ecosystems.

Runoff. That part of *precipitation*, snowmelt, or irrigation water that flows from the land to streams or other *surface waters*.

S

Salinity. The degree of salt in water.

Salt Water Intrusion. The invasion of fresh, surface, or groundwater by salt water.

Sand Dune. A ridge or hill of sand formed by wind action.

Sanitary Landfill. Landfill for disposing of solid wastes.

Sanitary Waste-water. Domestic waste-water without storm and surface runoff. Waste-water from the sanitary conveniences in dwellings, office buildings, industrial plants and institutions. Water supply of a community after it has been used and discharged to a sewer.

Saprophytes. Organisms living on dead or decaying organic matter that help natural decomposition of organic matter in water.

Saturated Zone. The area below the water table where all open spaces are filled with water under pressure equal to or greater than that of the atmosphere.

Savanna. An area with trees scattered widely among dense grasses.

Scoping. The process of early identification of important environmental issues that require detailed evaluation.

Seasonal Variation. The change in a set of meteorological parameters averaged over three months. Seasonal variation is the largest climatic variation, and temperature is the most frequently observed meteorological parameter. Often, monthly averaged data are grouped into seasons, according to the prescribed definition.

Sea Surface Temperature. The temperature of the layer of seawater (approximately 0.5 m deep) nearest the atmosphere.

Sea Surface Temperature Anomalies. Temperature of emitted energy from the sea surface. SST anomaly = (SST - SST mean).

Second Assessment Report (SAR). The Second Assessment Report, prepared by the Intergovernmental Panel on Climate Change, reviewed the existing scientific literature on climate change. Finalized in 1995, it is comprised of three volumes. Science; Impacts, Adaptations and Mitigation; and Economic and Social Dimensions of Climate Change.

Second Growth. Forest that has been clear-cut and regrown.

Secondary Oil Recovery. Injection of water into an oil well after primary oil recovery to force out some of the remaining thicker crude oil.

Secondary Pollutants. Air pollutants produced through reactions between primary pollutants and normal atmospheric compounds. An example is ozone that forms over urban areas through reactions of primary pollutants, sunlight and natural atmospheric gases.

Secondary Succession. The re-establishment of an ecosystem where there are remnants of a previous biological community.

Secondary Treatment (of waste-water). Use of biological processes to degrade waste-water in a treatment facility.

Sector. Division, most commonly used to denote type of energy consumer (*e.g.*, residential) or according to the Intergovernmental Panel on Climate Change, the type of greenhouse gas emitter (*e.g.* industrial process).

Secular Carbon Dioxide Trend. The fairly uniform and accelerating increase of carbon dioxide concentration in the atmosphere, as illustrated by the *Mauna Loa record.* The secular trend reflects the increase in global atmospheric carbon dioxide concentrations produced by combustion of fossil fuels, kilning of limestone, and possibly a net biospheric release of carbon dioxide resulting from deforestation.

Secure Landfill. A type of landfill designed specifically for hazardous waste. Similar to a modern sanitary landfill in that it includes multiple barriers and collection systems to ensure that leachate does not contaminate soil and other resources.

Sediment Pollution. By volume and mass, sediment is our greatest water pollutant. It may choke streams, fill reservoirs, bury vegetation and generally create a nuisance that is difficult to remove.

Sedimentation. Removal of settleable suspended solids from a water or waste-water by gravity settling in a quiescent tank or basin. Also called clarification or settling.

Sediments. Soil, sand, and minerals washed from land into water, usually after rain. They pile up in reservoirs, rivers and harbors, destroying fish and wildlife habitat, and clouding the water so that sunlight cannot reach aquatic plants. Careless farming, mining, and building activities will expose sediment materials, allowing them to wash off the land after rainfall.

Seed Tree Cutting. A logging method in which mature trees with good genetic characteristics and high seed production are preserved to promote regeneration of the forest. It is an alternative to clear-cutting.

Seepage. Percolation of water through the soil from unlined canals, ditches, laterals, watercourses, or water storage facilities.

Seismic. Referring to vibrations in the Earth produced by earthquakes.

Self-noise. The noise generated by air flowing through a silencer.

Sensible Heat. The excess radiative energy that has passed from the Earth's surface to *the atmosphere* through advection, conduction, and convection processes.

Septic Tank. Underground tank for treatment of waste-water from a home in rural and suburban areas. Bacteria in the tank decompose organic wastes and the sludge settles to the bottom of the tank. The effluent flows out of the tank into the ground through a field of drainpipes.

Sequestration. Opportunities to remove atmospheric CO_2, either through biological processes (*e.g.* plants and trees), or geological processes through storage of CO_2 in underground reservoirs.

Sewage Treatment (primary). Mechanical treatment of sewage in which large solids are filtered out by screens and suspended solids settle out as sludge in a sedimentation tank.

Shale Oil. Slow-flowing, dark brown, heavy oil obtained when kerogen in oil shale is vaporized at high temperatures and then condensed. Shale oil can be refined to yield gasoline, heating oil, and other petroleum products.

Short Ton. Common measurement for a ton in the United States. A short ton is equal to 2,000 lbs or 0.907 metric tons.

Shortwave Radiation. The radiation received from the sun and emitted in the spectral wavelengths less than 4´m. It is also called solar radiation.

Signal-to-noise Ratio. A quantitative measure of the statistical detectability of a signal, expressed as a ratio of the magnitude of the signal relative to the variability. For first detection of a CO_2-induced climate change, the model signal is the mean change or anomaly in some climatic variable, usually surface air temperature, attributed by a numerical model to increased concentrations of carbon dioxide. Observed noise is the standard deviation or natural variability computed from observations of that variable and adjusted for sample size, autocorrelation, and time averaging.

Silt. Sediment between 1/16 and 1/256 mm in diameter.

Silviculture. Management of forest land for timber.

Sink. A reservoir that uptakes a chemical element or compound from another part of its cycle. For example, soil and trees tend to act as natural sinks for carbon.

Sludge. Gooey solid mixture of bacteria and virus laden organic matter, toxic metals, synthetic organic chemicals, and solid chemicals removed from waste-water at a sewage treatment plant.

Sludge Processing. Collection, treatment, and disposal of sludge.

Smog. Air pollution associated with oxidants.

Smoke. Particles suspended in air after incomplete combustion of materials.

Soft Water. Any water that does not contain a significant amount of dissolved minerals such as salts of calcium or magnesium.

Soil. Complex mixture of inorganic minerals (*i.e.* mostly clay, silt, and sand), decaying organic matter, water, air, and living organisms.

Soil and Water Conservation Practices. Control measures consisting of managerial, vegetative, and structural practices to reduce the loss of soil and water.

Soil Carbon. A major component of the terrestrial biosphere pool in the carbon cycle. The amount of carbon in the soil is a function of the historical vegetative cover and productivity, which in turn is dependent in part upon climatic variables.

Soil Conditioner. An organic material like humus or compost that helps soil absorb water, build a bacterial community, and take up mineral nutrients.

Soil Erodibility. An indicator of a soil's susceptibility to raindrop impact, runoff, and other erosive processes.

Soil Fertility. The capacity of a soil to supply the nutrients and physical properties necessary for plant growth.

Soil Horizon. A layer in soil (A,B,C) that differs from another layer in chemical, physical and biological properties.

Soil Moisture. The water contained in the pore space of the unsaturated zone.

Solar Constant. The rate at which solar energy is received just outside the Earth's *atmosphere* on a surface that is normal to the incident radiation and at the mean distance of the Earth from the sun. The current value is 0.140 watt/cm^2.

Solar Cycle. The periodic change in *sunspot* numbers. It is the interval between successive minima and is about 11.1 years.

Solar Energy. Direct radiant energy from the sun. It also includes indirect forms of energy such as wind, falling or flowing water (hydropower), ocean thermal gradients, and biomass, which are produced when direct solar energy interact with the earth.

Solar Pond. Shallow pond filled with water and used to generate relatively low-temperature water.

Solar Radiation. Energy from the Sun. Also referred to as short-wave radiation. Of importance to the climate system, solar radiation includes ultraviolet radiation, visible radiation, and infrared radiation.

Solar Wind. A continuous plasma stream expanding into interplanetary space from the sun's corona. The solar wind is present continuously in interplanetary space. After escaping from the gravitational field of the sun, this gas flows outward at a typical speed of 400 km per second to distances known to be beyond the orbit of Pluto. Besides affecting Earth's weather solar activity gives rise to a dramatic visual phenomena in our atmosphere.

Solid Waste. Non-liquid, non-soluble materials ranging from municipal garbage to industrial wastes that contain complex and sometimes hazardous substances. Solid wastes also include sewage sludge, agricultural refuse, demolition wastes, and mining residues. Technically, solid waste also refers to liquids and gases in containers.

Soot. Carbon dust formed by incomplete combustion.

Source. Any process or activity that releases a greenhouse gas, an aerosol, or a precursor of a greenhouse gas into the atmosphere.

Southern Oscillation. A large-scale atmospheric and hydrospheric fluctuation centered in the equatorial Pacific Ocean. It exhibits a nearly annual pressure anomaly, alternatively high over the Indian Ocean and high over the South Pacific. Its period is slightly variable, averaging 2.33 years. The variation in pressure is accompanied by variations in wind strengths, ocean currents, sea-surface temperatures, and precipitation in the surrounding areas. *El Nino* occurrences are associated with the phenomenon.

Southern Oscillation Index. An indicator based on the pressure gradient between the quasi-stationary low pressure region and the center of a subtropical high pressure cell. A positive index corresponds to an anomalously high pressure difference between the two centers of action.

Special Naphtha. All finished products within the naphtha boiling range that are used as paint thinners, cleaners, or solvents. Those products are refined to a specified flash point.

Species. A group of individuals capable of interbreeding.

Spring. Ground water seeping out of the earth where the water table intersects the ground surface.

Spring Melt/Thaw. The process whereby warm temperatures melt winter snow and ice. Because various forms of acid deposition may have been stored in the frozen water, the melt can result in abnormally large amounts of acidity entering streams and rivers, sometimes causing fish kills.

SRES Scenarios. A suite of emissions scenarios developed by the Intergovernmental Panel on Climate Change in its Special Report on Emissions Scenarios (SRES). These scenarios were developed to explore a range of potential future greenhouse gas emissions pathways over the 21st century and their subsequent implications for global climate change.

Stable Air. A motionless mass of air that holds, instead of dispersing, pollutants.

Stagnation. Lack of motion in a mass of air or water that holds pollutants in place.

Stationary Sources. Air pollution sources that have a relatively fixed location, including point sources, fugitive sources and area sources.

Statistical-dynamical Models. Computer programs that calculate simplified climate models based on versions of the conservation equations that have been averaged over longitude, with the effects of the synoptic eddies parameterized statistically in the meridional plane.

Steric Height. The mean dynamic depth (or height) of the ocean for the month minus the annual mean dynamic depth for the same isobaric reference level.

Still Gas. Any form or mixture of gases produced in refineries by distillation, cracking, reforming, and other processes. Principal constituents are methane, ethane, ethylene, normal butane, butylene, propane, propylene, etc. Used as a refinery fuel and as a petrochemical feedstock.

Stoma, Plant Stomata. A minute pore in the epidermis of plant leaves or stems. Stoma, which are bordered by guard cells that regulate the size of the opening, function in gas exchange between the plant and the external environment. The stomatal apparatus or stomate consists of the stoma plus guard cells.

Stratification. Separating into layers.

Stratosphere. Second layer of the atmosphere, extending from about 19 to 48 kilometers (12 to 30 miles) above the earth's surface. It contains small amounts of gaseous ozone (O_3), which filters out about 99 per cent of the incoming harmful ultraviolet (UV) radiation. Most commercial airline flights operate at a cruising altitude in the lower stratosphere.

Stratospheric Ozone. See *ozone layer*.

Strip-Cropping. Growing crops in a systematic arrangement of strips or bands that serve as barriers to wind and water erosion.

Strip Mining. Cutting deep trenches to remove minerals such as coal and phosphate found near the earth's surface in flat or rolling terrain.

Subbituminous Coal. A dull, black coal of rank intermediate between lignite and bituminous coal.

Submerged Aquatic Vegetation. Vegetation that lives at or below the water surface; an important habitat for young fish and other aquatic organisms.

Subsidence. A thinking, settling, or otherwise lowering of parts of the crust of the Earth.

Subsidiary Body for Implementation (SBI). A permanent body established by the UNFCCC that makes recommendations to the COP on policy and implementation issues. It is open to participation by all Parties and is composed of government representatives.

Subsidiary Body for Scientific and Technological Advice (SBSTA). A permanent body established by the UNFCCC that serves as a link between expert information sources such as the IPCC and the COP.

Subwatershed. Topographic perimeter of the catchment area of a stream tributary.

Suess Effect. The relative change in the 14C/C or 13C/C ratio of any carbon pool or reservoir caused by the addition of fossil-fuel CO_2 to the atmosphere. Fossil fuels are devoid of 14C because of the radioactive decay of 14C to 14N during long underground storage and are depleted in 13C because of isotopic fractionation eons ago during photosynthesis by the plants that were the precursors of the fossil fuels. Carbon dioxide produced by the combustion of fossil fuels is thus virtually free of 14C and depleted in 13C. The term Suess effect originally referred to the dilution of the 14C/C ratio in atmospheric CO_2 by the admixture of fossil-fuel produced CO_2, but the definition has been extended to both the 14C and 13C ratios in any pool or reservoir of the carbon cycle resulting from human disturbances.

Sulfate Aerosols. Particulate matter that consists of compounds of sulfur formed by the interaction of sulfur dioxide and sulfur trioxide with other compounds in the atmosphere. Sulfate aerosols are injected into the atmosphere from the combustion of fossil fuels and the eruption of volcanoes like Mt. Pinatubo. Recent theory suggests that sulfate aerosols may lower the earth's temperature by reflecting away solar radiation (negative radiative forcing). General Circulation Models which incorporate the effects of sulfate aerosols more accurately predict global temperature variations.

Sulfur Cycle. Cyclic movement of sulfur in different chemical forms from the environment, to organisms, and then back to the environment.

Sulfur Dioxide (SO_2). A compound composed of one sulfur and two oxygen molecules. Sulfur dioxide emitted into the atmosphere through natural and anthropogenic processes is changed in a complex series of chemical reactions in the atmosphere to sulfate aerosols. These aerosols are believed to result in negative radiative forcing (*i.e.*, tending to cool the Earth's surface) and do result in acid deposition (*e.g.*, acid rain).

Sulfur Hexafluoride (SF_6). A colorless gas soluble in alcohol and ether, slightly soluble in water. A very powerful greenhouse gas used primarily in electrical transmission and distribution systems and as a dielectric in electronics. The global warming potential of SF_6 is 23,900.

Sulphurous Smog. Produced primarily by burning coal or oil at large power plants. Sulphur oxides and particulates combine under certain meteorological conditions to produce a concentrated form of this smog.

Sump. A pit or tank that catches liquid runoff for drainage or disposal.

Sun. The closest star to Earth (149,599,000 km away on average). The sun dwarfs the other bodies in the solar system, representing approximately 99.86 per cent of all the mass in the solar system. One hundred and nine Earths would be required to fit across the Sun's disk, its interior could hold over 1.3 million Earths.

Sunspot. A relatively dark, sharply defined region on the solar disk, marked by an umbra approximately 2000K cooler than the effective photospheric temperature, surrounded by a less dark but also sharply bounded penumbra. The average spot diameter is about 3700 km, but can range up to 245,000 km. Most sunspots are found in groups of two or more, but they can occur singly. Sunspots are cyclic, with a period of approximately 11 years. The quantitative description of sunspot activity is called the Wolf sunspot number, denoted R. The Wolf sunspot number is also referred to as Wolfer sunspot number, Zurich relative sunspot number, or relative sunspot number.

Supplementarity. The Protocol does not allow Annex I parties to meet their emission targets entirely through use of emissions trading and the other Kyoto Mechanisms; use of the mechanisms must be supplemental to domestic actions to limit or reduce their emissions.

Surface Air Temperature. The temperature of the air near the surface of the Earth, usually determined by a thermometer in an instrument shelter about 2 m above the ground. The true daily mean, obtained from a thermograph, is approximated by the mean of 24 hourly readings and may differ by 1.0 degrees C from the average based on minimum and maximum readings. The global average surface air temperature is 15 degrees C.

Surface Albedo. The fraction of solar radiation incident on the Earth's surface that is reflected by it. Reflectivity varies with ground cover, and during the winter months it varies greatly with the amount of snow cover (depth and areal extent). Roughness of terrain, moisture content, solar angle, and angular and spectral distribution of ground-level irradiations are other factors affecting surface albedo.

Surface Impoundment. Treatment, storage, or disposal of liquid hazardous wastes in ponds.

Surface Mining. Removal of soil, sub-soil, and other strata and then extracting a mineral deposit found fairly close to the earth's surface.

Surface Runoff. Precipitation, snow melt, or irrigation water in excess of what can infiltrate the soil surface and be stored in small surface depressions; a major transporter of non-point source pollutants in rivers, streams, and lakes.

Surface Water. All water naturally open to the atmosphere.

Suspended Loads. Specific sediment particles maintained in the water column by turbulence and carried with the flow of water.

Suspended Solids. Small particles of solid pollutants that float on the surface of, or are suspended in, sewage or other liquids. They resist removal by conventional means.

Sustainability. A concept that is emerging in the environmental sciences. With respect to resources it involves management that has the objective of ensuring that future generations will have the opportunity to use their fair share of resources and will inherit a quality environment. In an economic sense the concept means development that will not cause irreparable damage to the environment while ensuring that future generations will inherit their fair share of all Earth's resources.

Sustainable Ecosystem. An ecosystem that is subject to some human use, but at a level that leads to no loss of species or of necessary ecosystem functions.

Sustainable Energy. A type of energy management that provides for reliable sources of energy while not causing environmental degradation and ensuring that future generations will have a fair share of the Earth's resources.

Sustainable Forest Ecosystem. An ecosystem in which all properties of the forest are maintained.

Sustainable Harvest. An amount of a resource that can be harvested at regular intervals indefinitely. There are two concepts here. The first is the maximum abundance of a population or species that can be harvested without degrading the ability of that population or species to withstand the same level of harvest in the future. The second concept extends sustainability to the ecosystem. It is maximum abundance of a population or species that can be harvested without degrading (i) the ability of that population or species; and (ii) its ecosystem to withstand the same level of harvest in the future.

Sustainable Water Use. Use of water resources that does not harm the environment and provides for the existence of high-quality water for future generations.

Swamp. A type of *wetland* that is dominated by woody vegetation and does not accumulate appreciable peat deposits; it may be fresh- or saltwater, and tidal or nontidal.

Synthetic Fertilizer. Commercially prepared mixtures of plant nutrients such as nitrates, phosphates, and potassium applied to the soil to restore fertility and increase crop yields.

Synthetic Natural Gas (SNG). A manufactured product chemically similar in most respects to natural gas, resulting from the conversion or reforming of petroleum hydrocarbons. It may easily be substituted for, or interchanged with, pipeline quality natural gas.

T

Tailings. Rock and other waste materials removed as impurities when minerals are mined and mineral deposits are processed. These materials are usually dumped on the ground or into ponds.

Tar Sand. Swamp-like deposit of a mixture of fine clay, sand, water, and variable amounts of tar-like heavy oil known as bitumen. Bitumen can be extracted from tar sand by heating. It can then be purified and upgraded to synthetic crude oil.

Targets and Timetables. Targets refer to the emission levels or emission rates set as goals for countries, sectors, companies, or facilities. When these goals are to be reached by specified years, the years at which goals are to be met are referred to as the timetables. In the Kyoto Protocol, a target is the percent reduction from the 1990 emissions baseline that the country has agreed to. On average, developed countries agreed to reduce emissions by 5.2% below 1990 emissions during the period 2008-2012, the first commitment period.

Taxon. A grouping of organisms according to evolutionary relationships.

Tectonic Cycle. The processes that change Earth's crust, producing external forms such as ocean basins, continents and mountains.

Temperature. Measure of the average speed of motion of the atoms or molecules in a substance or combination of substances at a given moment.

Tephra. Any rock material produced by a volcano.

Terracing. Dikes built along the contour of sloping farm land that hold runoff and sediment to reduce erosion.

Terrestrial. Pertaining to land.

Terrestrial Radiation. The total infrared radiation emitted by the Earth and its atmosphere in the temperature range of approximately 200 to 300 Kelvin. Terrestrial radiation provides a major part of the potential energy changes necessary to drive the atmospheric wind system and is responsible for maintaining the surface air temperature within limits of livability.

Tertiary Treatment (of waste-water). Advanced form of waste-water treatment involving chemical treatment or advanced filtration. An example is chlorination of water.

Thermal Expansion. Expansion of a substance as a result of the addition of heat. In the context of climate change, thermal expansion of the world's oceans in response to global warming is considered the predominant driver of current and future sea-level rise.

Thermal Pollution. Discharge of heated water from industrial processes that can kill or injure aquatic organisms.

Thermal Stratification. The formation of layers of different temperatures in a lake or reservoir.

Thermocline. A transition layer of water in the ocean, with a steeper vertical temperature gradient than that found in the layers of ocean above and below. The permanent thermocline separates the warm mixed surface layer of the ocean from the cold deep ocean water, and is found between 100- and 1000-m depths. The thermocline first appears at the 55-60 degree N and S latitudes, where it forms a horizontal separation between temperate and polar waters. The thermocline reaches its maximum depth at mid-latitudes and is shallowest at the equator and at its northern and southern limits. The thermocline is stably stratified, and transfer of water and carbon dioxide across this zone occurs very slowly. Thus, the thermocline acts as a barrier to the downward mixing of carbon dioxide.

Thermohaline. Refers to the combined effects of temperature and salinity that contribute to density variations in the oceans.

Thermohaline Circulation (THC). A three-dimensional pattern of ocean circulation driven by wind, heat and salinity that is an important component of the ocean-atmosphere climate system. In the Atlantic, winds transport warm tropical surface water northward where it cools, becomes more dense, and sinks into the deep ocean, at which point it reverses direction and migrates back to the tropics, where it eventually warms and returns to the surface. This cycle or "conveyor belt" is a major mechanism for the global transport of heat, and thus has an important influence on the climate. Global warming is projected to increase sea-surface temperatures, which may slow the THC by reducing the sinking of cold water in the North Atlantic. In addition, ocean salinity also influences water density, and thus decreases in sea-surface salinity from the melting of ice caps and glaciers may also slow the THC.

Thunder. The sound that results from lightning. Lightning bolts (static electricity) produce intense heat. This burst of heat makes the air around the bolt expand explosively producing the sound we hear as thunder. Since light travels faster than sound, we see the lightning before we hear the thunder.

Thunderstorm. Local storm resulting from warm humid air rising in an unstable environment. Air may start moving upward because of unequal surface heating, the lifting of warm air along a frontal zone, or diverging upper-level winds (these diverging winds draw air up beneath them). The scattered thunderstorms that develop in the summer are called airmass thunderstorms because they form in warm, maritime tropical air masses away from other weather fronts. More violent severe thunderstorms form in areas with a strong vertical wind shear that forces the updraft into the mature stage, the most intense stage of the thunderstorm. Severe thunderstorms can produce large hail, forceful winds, flash floods, and tornadoes.

Tidal Marsh. Low, flat marshlands traversed by channels and tidal hollows and subject to tidal innundation; normally, the only vegetation present are salt-tolerant bushes and grasses.

Tidal Power. Form of water utilising ocean tides in places where favourable topography allows for construction of a power plant.

TOC. Total organic carbon—a measure of the amount of organic carbon in water.

Tornado. A twisting, spinning funnel of low pressure air. The most unpredictable weather event, tornadoes are created during powerful thunderstorms. As a column of warm air rises, air rushes in at ground level and begins to spin. If the storm gathers energy a twisting, spinning funnel develops. Because of the funnel's cloud and rain composition and the dust, soil, and debris it draws up, the funnel appears blackish in color. The most energetic storms result in the funnel touching the ground. In these tornadoes, the roaring winds in the funnel can reach 300 mph, the strongest winds on Earth. Funnels usually travel at 20 to 40 mph, moving toward the northeast. When tornadoes form over lakes or oceans they suck water into the funnel cloud and are called waterspouts.

Toxic. Harmful, deadly, or poisonous.

Toxic Cloud. Airborne plume of gases, vapors, fumes, or aerosols containing toxic materials.

Toxic Pollutant. Those pollutants, or combinations of pollutants, including disease causing agents, which after discharge and upon exposure, ingestion, inhalation or assimilation into any organism, either directly from the environment or indirectly through food chains, will cause death, disease, behavioural abnormalities, cancer, genetic mutations, physiological malfunction (including malfunctions in reproduction) or physical deformations, in such organisms or their offspring.

Toxicology. The science concerned with study of poisons (or toxins) and their effects on living organisms. The subject also includes the clinical, industrial, economic and legal problems associated with toxic materials.

Trace Gas. A minor constituent of the atmosphere. The most important trace gases contributing to the greenhouse effect are water vapor, carbon dioxide, ozone, methane, ammonia, nitric acid, nitrous oxide, ethylene, sulfur dioxide, nitric oxide, dichlorofluoromethane or Freon 12, trichlorofluoromethane or Freon 11, methyl chloride, carbon monoxide, and carbon tetrachloride.

Trade Winds. Surface air from the horse latitudes that moves back toward the equator and is deflected by the Coriolis Force, causing the winds to blow from the Northeast in the Northern Hemisphere and from the Southeast in the Southern Hemisphere. These steady winds are called trade winds because they provided trade ships with an ocean route to the New World.

Transboundary Pollutants. Air pollution that travels from one jurisdiction to another, often crossing state or international boundaries. Also applies to water pollution.

Transient Tracers. Chemical elements (often radioactive) or compounds that have finite lifetimes. Atmospheric testing of nuclear weapons from the mid-1950s to the early 1960s released large quantities of radionuclides to the atmosphere. Atmosphere-ocean exchange processes have transferred some of these elements to the oceans. Studying the behavior and distribution of these specific isotopes and other chemical tracers in the ocean will provide information on: residence times of the water and its dissolved components in gyres, basins, etc.; the mode and rate of formation and the subsequent spreading rates of specific water types, such as the polar water of the Norwegian and Greenland Seas; deep-ocean circulation and ocean-mixing processes, such as advection and upwelling; and the flux of anthropogenic carbon dioxide into the ocean through its correlation with several different transient tracers.

Transpiration. The process in plants by which water is taken up by the roots and released as water vapor by the leaves. The term can also be applied to the quantity of water thus dissipated.

Transportation Sector. Consists of private and public passenger and freight transportation, as well as government transportation, including military operations.

Trash. Material considered worthless or offensive that is thrown away. Generally defined as dry waste material, but in common usage it is a synonym for garbage, rubbish, or refuse.

Trash-to-Energy Plan. Burning trash to produce energy.

Treated Waste-water. Waste-water that has been subjected to one or more physical, chemical, and biological processes to reduce its potential of being health hazard.

Treatment. • Any method, technique, or process designed to remove solids and/or pollutants from solid waste, waste-streams, effluents, and air emissions. • Methods used to change the biological character or composition of any regulated medical waste so as to substantially reduce or eliminate its potential for causing disease.

Treatment Plant. A structure built to treat waste-water before discharging it into the environment.

Tree Rings. Annual growth increments of trees that indicate, among other factors, the climatic conditions that enhance or limit growth. Tree ring widths and indexes have been used to search for solar-terrestrial relationships and climatic cycles and to reconstruct past climates.

Trophic Level. A segment of the food chain in which all organisms obtain food and energy in, basically, the same manner (*e.g.*, *photosynthesis*, herbivory, or carnivory) and in which all organisms are the same number of links from the photosynthetic segment.

Tropical Ocean-Global Atmosphere (TOGA). TOGA is a program jointly sponsored by the United Nations World Meteorological Organization (WMO); the International Council of Scientific Unions (ICSU); the United Nations Educational, Scientific, and Cultural Organization (UNESCO) Intergovernmental Oceanographic Commission (IOC); and the ICSU Scientific Committee on Oceanic Research (SCOR).

TOGA has four major objectives:

- To collect and catalog observations of the tropical atmosphere and ocean;
- To assess the evolution of the tropical atmosphere/ocean system in real time;
- To promote the development of short-term climate-prediction computer models for the tropics;
- To study the influence of the tropical atmosphere/ocean system on the climate at higher latitudes.

Tropical Storm Formation. Tropical storms generally form in the eastern portion of tropical oceans and track westward. Hurricanes, typhoons, and willywuhes all start out as weak low pressure areas that form over warm tropical waters (*e.g.*, surface water temperature of at least 80° F). Initially winds and cloud formations over the warm tropical waters are minimal. Both intensify with time. Formation of tropical storms also requires a significant Coriolis effect to induce proper spin in the wind formation. As the storm begins to organize itself into a coherent pattern, it will experience increased activity and intensity.

When a storm develops a clearly recognizable pattern, it is referred to as a tropical depression. When wind speeds reach 35 knots (40.3 mph), it is called a tropical storm and is given a name. When wind speed equals or exceeds 74 mph, the storm is called a hurricane. In the western Pacific, a hurricane is referred to as a typhoon. In waters around Australia it is called a cyclone or willy-willy.

Hurricanes intensify when moving over areas of increased water temperatures, and weaken over colder water surfaces. Upper atmosphere wind shear (different wind direction and speeds at different elevations) will frequently prevent or slow intensification of tropical storms by "spreading out" the storm horizontally and preventing the formation of strong updrafts of warm, humid air. Movement over a land-mass will weaken hurricane winds but will result in large-scale rain that can result in large-scale flooding. When encountering a strong frontal system (such as a polar front) the hurricane will curve and track along the leading edge of the front or become implanted in it.

Tropopause. The boundary between the troposphere and the stratosphere (about 8 km in polar regions and about 15 km in tropical regions), usually characterized by an abrupt change of *lapse rate*. The regions above the troposphere have increased atmospheric stability than those below. The tropopause marks the vertical limit of most clouds and storms.

Troposphere. The lowest layer of the atmosphere and contains about 95 per cent of the mass of air in the Earth's atmosphere. The troposphere extends from the Earth's surface up to about 10 to 15 kilometers. All weather processes take place in the troposphere. Ozone that is formed in the troposphere plays a significant role in both the greenhouse gas effect and urban smog.

Tropospheric Ozone (O_3). See *ozone*.

Tropospheric Ozone Precursors. See *ozone precursors*.

Trough. Elongated area of low atmospheric pressure, either at the surface or in the upper atmosphere.

Tundra. A type of *ecosystem* dominated by lichens, mosses, grasses, and woody plants. It is found at high latitudes (arctic tundra) and high altitudes (alpine tundra). Arctic tundra is underlain by *permafrost* and usually very wet.

Turbidimeter. A device that measures the cloudiness of suspended solids in a liquid; a measure of the quantity of suspended solids.

Turbidity. 1. Haziness in air caused by the presence of particles and pollutants. 2. A cloudy condition in water due to suspended silt or organic matter.

Turnover Rate. The fraction of the total amount of mass (*e.g.*, carbon) in a given pool or reservoir that is released from or that enters the pool in a given length of time. The turnover rate of carbon is often expressed as GtC/year.

Typhoon. Hurricanes in the Western Pacific Ocean.

U

Ubiquitous Species. Species that are found almost any where on the Earth.

Ultraviolet Radiation (UV). A portion of the electromagnetic spectrum with wavelengths shorted than visible light. The sun produces UV, which is commonly split into three bands of decreasing wavelength. Shorter wavelength radiation has a greater potential to cause biological damage on living organisms. The longer wavelength ultraviolet band, UVA, is not absorbed by ozone in the atmosphere. UVB is mostly absorbed by ozone, although some reaches the Earth. The shortest wavelength band, UVC, is completely absorbed by ozone and normal oxygen in the atmosphere.

Ultraviolet Rays. Radiation from the sun that can be useful or potentially harmful. UV rays from one part of the spectrum (UV-A) enhance plant life. UV rays from other parts of the spectrum (UV-B) can cause skin cancer or other tissue damage. The ozone layer in the atmosphere partly shields us from ultraviolet rays reaching the earth's surface.

Umbrella Group. Negotiating group within the UNFCCC process comprising the United States, Canada, Japan, Australia, New Zealand, Norway, Iceland, Russia, and Ukraine.

Underground Storage Tank (UST). A tank located at least partially underground and designed to hold gasoline or other petroleum products or chemicals.

Uneven-aged Stands. Forest area with at least three distinct age classes.

Unfinished Oils. All oils requiring further refinery processing, except those requiring only mechanical blending. Includes naphtha and lighter oils, kerosene and light gas oils, heavy gas oils, and residuum.

Unified Soil Classification System. A classification of soils, widely used in engineering practice, based on the amount of coarse particles, fine particles, or organic material.

United Nations Framework Convention on Climate Change. The international treaty unveiled at the United Nations Conference on Environment and Development (UNCED) in June 1992. The UNFCCC commits signatory countries to stabilize anthropogenic (*i.e.* human-induced) greenhouse gas emissions to "levels that would prevent dangerous anthropogenic interference with the climate system." The UNFCCC also requires that all signatory parties develop and update national inventories of anthropogenic emissions of all greenhouse gases not otherwise controlled by the Montreal Protocol. Out of 155 countries that have ratified this accord, the United States was the first industrialized nation to do so.

Unsaturated Zone. The area above the water table where soil pores are not fully saturated, although some water may be present.

Upwelling. The vertical motion of water in the ocean by which subsurface water of lower temperature and greater density moves toward the surface of the ocean. Upwelling occurs most commonly among the western coastlines of continents, but may occur anywhere in the ocean. Upwelling results when winds blowing nearly parallel to a continental coastline transport the light surface water away from the coast. Subsurface water of greater density and lower temperature replaces the surface water, and exerts a considerable influence on the weather of coastal regions. Carbon dioxide is transferred to the atmosphere in regions of upwelling. This is especially important in the Pacific equatorial regions, where 1 - 2 GtC/year may be released to the atmosphere. Upwelling also results in increased ocean productivity by transporting nutrient-rich waters to the surface layer of the ocean.

Urban Dust Dome. Polluted urban air produced by the combination of lingering air and abundance of particulates and other pollutants in the urban air mass.

Urban Forestry. The practice and profession of planting and maintaining trees in cities, including trees in parks and other public areas. Involves determining the best species and sites for urban tree planting, taking into account climate, soil, shading from tall buildings and motor vehicle pollution.

Urban Heat Island (UHI). Refers to the tendency for urban areas to have warmer air temperatures than the surrounding rural landscape, due to the low albedo of streets, sidewalks, parking lots, and buildings. These surfaces absorb solar radiation during the day and release it at night, resulting in higher night temperatures.

Urban Runoff. Storm water from city streets and adjacent domestic or commercial properties that carries pollutants of various kinds into the sewer systems and receiving waters.

V

Vadose Zone. The zone between land surface and the water table within which the moisture content is less than saturation (except in the capillary fringe) and pressure is less than atmospheric. Soil pore space also typically contains air or other gases. The capillary fringe is included in the vadose zone.

Valued Environmental Attributes/Components. Those aspects(components/processes/functions) of ecosystems, human health, and environmental welfare considered to be important and potentially at risk from human activity or natural hazards. Similar to the term "valued environmental components" used in environmental impact assessment.

Vapor. The gaseous phase of substances that are liquid or solid at atmospheric pressure (*e.g.*, steam).

Vapor Dispersion. The movement of vapor clouds in air due to wind, thermal action, gravity spreading, and mixing.

Vapor Plumes. Flue gases visible because they contain water droplets.

Vapor Pressure. A measure of a substance's propensity to evaporate, vapor pressure is the force per unit area exerted by vapor in an equilibrium state with surroundings at a given pressure. It increases exponentially with an increase in temperature. A relative measure of chemical volatility, vapor pressure is used to calculate water partition coefficients and volatilization rate constants.

Vapor Recovery System. A system by which the volatile gases from gasoline are captured instead of being released into the atmosphere.

Varve. A layer of sediment deposited in lakes during one year. Each layer consists of two parts, which are deposited at different seasons and which differ in color and texture; thus, the layers can be counted and measured. In a complete series, the number of layers gives the date on which the ground was vacated by the retreating ice.

Vector. • An organism, often an insect or rodent, that carries disease. • Plasmids, viruses, or bacteria used to transport genes into a host cell. A gene is placed in the vector; the vector then "infects" the bacterium.

Vector-borne Disease. Disease that results from an infection transmitted to humans and other animals by blood-feeding anthropods, such as mosquitoes, ticks, and fleas. Examples of vector-borne diseases include Dengue fever, viral encephalitis, Lyme disease, and malaria.

Vegetative Controls. Non-point source pollution control practices that involve vegetative cover to reduce erosion and minimize loss of pollutants.

Vehicle Miles Traveled (VMT). One vehicle traveling the distance of one mile. Thus, total vehicle miles is the total mileage traveled by all vehicles.

Ventilation Rate. The rate at which indoor air enters and leaves a building. Expressed as the number of changes of outdoor air per unit of time (air changes per hour (ACH), or the rate at which a volume of outdoor air enters in cubic feet per minute (CFM).

Ventilation/Suction. The act of admitting fresh air into a space in order to replace stale or contaminated air; achieved by blowing air into the space. Similarly, suction represents the admission of fresh air into an interior space by lowering the pressure outside of the space, thereby drawing the contaminated air outward.

Venturi Scrubbers. Air pollution control devices that use water to remove particulate matter from emissions.

Vernal Equinox. The beginning of spring in the Northern Hemisphere. The time/day that the sun crosses the equatorial plane going from south to north.

Virgin Forest. Forest that has never been cut.

Virgin Materials. Resources extracted from nature in their raw form, such as timber or metal ore.

Virus. Any of a large group of submicroscopic infective agents that typically contain a protein coat surrounding a nucleic acid core and are capable of growth only in a living cell.

Volatile Organic Compounds (VOCs). Organic compounds that evaporate readily into the atmosphere at normal temperatures. VOCs contribute significantly to photochemical smog production and certain health problems.

Volatile Synthetic Organic Chemicals. Chemicals that tend to volatilize or evaporate.

Volcano. A naturally occurring vent or fissure at the Earth's surface through which erupt molten, solid, and gaseous materials. Volcanic eruptions inject large quantities of dust, gas, and aerosols into the atmosphere. A major component of volcanic clouds is sulfur dioxide, a strong absorber of ultraviolet radiation. Chemical interactions between sulfur dioxide and water cause sulfuric acid aerosols which can scatter some of the incident solar radiation back to space, thus causing a global cooling effect.

Volume Reduction. Processing waste materials to decrease the amount of space they occupy, usually by compacting, shredding, incineration, or composting.

Vulnerable Species. Another term for threatened species—species experiencing a decline in the number of individuals.

Vulnerable Zone. An area over which the airborne concentration of a chemical accidentally released could reach the level of concern.

W

Waldsterben. German phenomenon of forest death as the result of acid rain, ozone and other air pollutants.

Walker Cell. A zonal circulation of the atmosphere confined to equatorial regions and driven principally by the oceanic temperature gradient. In the Pacific, air flows westward from the colder, eastern area to the warm, western ocean, where it acquires warmth and moisture and subsequently rises. A return flow aloft and subsidence over the eastern ocean complete the cell.

Waste. • Unwanted materials left over from a manufacturing process. • Refuse from places of human or animal habitation.

Waste Characterization. Identification of chemical and microbiological constituents of a waste material.

Waste Exchange. Arrangement in which companies exchange their wastes for the benefit of both parties.

Waste Feed. The continuous or intermittent flow of wastes into an incinerator.

Waste Generation. The weight or volume of materials and products that enter the waste stream before recycling, composting, landfilling, or combustion takes place. Also can represent the amount of waste generated by a given source or category of sources.

Waste Load Allocation. (1) The maximum load of pollutants each discharger of waste is allowed to release into a particular waterway. Discharge limits are usually required for each specific water quality criterion being, or expected to be violated. (2) The portion of a stream's total assimilative capacity assigned to an individual discharge.

Waste Minimization. Measures or techniques that reduce the amount of wastes generated during industrial production processes; term is also applied to recycling and other efforts to reduce the amount of waste going into the waste stream.

Waste Reduction. Using source reduction, recycling, or composting to prevent or reduce waste generation.

Waste Stream. The total flow of solid waste from homes, businesses, institutions, and manufacturing plants that is recycled, burned, or disposed of in landfills, or segments thereof such as the "residential waste stream" or the "recyclable waste stream."

Waste Treatment Lagoon. Impoundment made by excavation or earth fill for biological treatment of waste-water.

Waste Treatment Plant. A facility containing a series of tanks, screens, filters and other processes by which pollutants are removed from water.

Waste-to-Energy Facility. Facility where recovered municipal solid waste is converted into a usable form of energy, usually via combustion.

Waste-water. Water that has been used and contains dissolved or suspended waste materials.

Waste-water Treatment Plan. A facility containing a series of tanks, screens, filters, and other processes by which pollutants are removed from water. Most treatments include chlorination to attain safe drinking water standards.

Water Solubility. The maximum possible concentration of a chemical compound dissolved in water. If a substance is water soluble it can very readily disperse through the environment.

Water Stress Effect. The closing of the *stomata* by a plant in response to excessive water loss through *transpiration* or in response to drought conditions. The stomatal closing reduces CO_2 uptake as well as water loss, thus decreasing the photosynthetic rate. However, under conditions of elevated CO_2 concentration, the CO_2 gradient between *the* *atmosphere* and the leaf is higher than under ambient conditions, and CO_2 can pass through partially closed stomates at a rate similar to that under conditions of lower CO_2 and open stomates. The humidity gradient remains the same at higher CO_2, and *transpiration* is impeded. The net result is improved water-use efficiency by some plants.

Water Table. The level of groundwater.

Water Treatment Lagoon. An impound for liquid wastes designed to accomplish some degree of biochemical treatment.

Water-use Efficiency. A measure of the amount of water used by plants per unit of plant material produced. The term can be applied at the leaf, whole-plant, and ecosystem levels. At the leaf level, it is more precisely referred to as the *instantaneous transpiration efficiency,* the CO_2 assimilation rate (*photosynthesis*) divided by the transpiration rate (the moles of CO_2 taken up divided by the moles of water lost through transpiration in a unit of time per unit leaf area). At the whole-plant level, it is more precisely referred to as the *growth water-use efficiency,* the units of dry matter synthesized divided by the units of water lost. At the ecosystem level, it is more precisely referred to as the *crop water-use efficiency,* the grams of dry weight gained by plants during the growing season per unit land area divided by the millimeters of water lost (including evaporation directly from the soil).

Water Vapor. The most abundant greenhouse gas; it is the water present in the atmosphere in gaseous form. Water vapor is an important part of the natural greenhouse effect. While humans are not significantly increasing its concentration, it contributes to the enhanced greenhouse effect because the warming influence of greenhouse gases leads to a positive water vapor feedback. In addition to its role as a natural greenhouse gas, water vapor plays an important role in regulating the temperature of the planet because clouds form when excess water vapor in the atmosphere condenses to form ice and water droplets and precipitation.

Water Vapor Feedback. A process in which an increase in the amount of water vapor increases *the atmosphere's* absorption of longwave radiation, thereby contributing to a warming of the atmosphere. Warming, in turn, may result in increased evaporation and an increase in the initial water vapor anomaly. This feedback, along with carbon dioxide, is responsible for the *greenhouse effect* and operates virtually continuously in the atmosphere.

Waterborne Disease Outbreak. The significant occurence of acute illness associated with drinking water from a public water system that is deficient in treatment, as determined by appropriate local or state agencies.

Watershed. The land area that drains into a stream; the watershed for a major river may encompass a number of smaller watersheds that ultimately combine at a common point.

Watershed Approach. A coordinated framework for environmental management that focuses public and private efforts on the highest priority problems within hydrologically-defined geographic areas taking into consideration both ground and surface water flow.

Watershed Area. A topographic area within a line drawn connecting the highest points uphill of a drinking waterintake into which overland flow drains.

Waxes. Solid or semisolid materials derived from petroleum distillates or residues. Light-colored, more or less translucent crystalline masses, slightly greasy to the touch, consisting of a mixture of solid hydrocarbons in which the paraffin series predominates. Included are all marketable waxes, whether crude scale or fully refined. Used primarily as industrial coating for surface protection.

Weather. Weather is the specific condition of the atmosphere at a particular place and time. It is measured in terms of such things as wind, temperature, humidity, atmospheric pressure, cloudiness, and precipitation. In most places, weather can change from hour-to-hour, day-to-day, and season-to-season. Climate is the average of weather over time and space. A simple way of remembering the difference is that climate is what you expect (*e.g.* cold winters) and 'weather' is what you get (*e.g.* a blizzard).

Weather Symbols. Some commonly used symbols are illustrated in the chart on the right.

Weather Terms.

- **Clear.** Sky cloud-free to 30 per cent covered.
- **Sunny.** Sunshine 70-100 per cent of the day.
- **Partly sunny** and **partly cloudy.** Both terms refer to 40 to 70 per cent cloud cover. Partly sunny is used in the day; partly cloudy is used at night.
- **Fog.** A cloud on the ground. Fog is composed of billions of tiny water droplets floating in the air.
- **Snow.** Precipitation of ice crystals.
- **Snow flurries.** Intermittent snowfall that may result in little accumulation.
- **Sleet.** Pellets of ice that form when rain or melting snowflakes freeze while falling. (Occurs in cold weather; hail usually occurs in summer.)
- **Freezing rain.** Rain that turns to ice on impact with the surface.
- **Rain.** Extended period of precipitation. Associated with large storm systems rather than single clouds or thunder storms.
- **Showers.** Brief interval of rain that does not affect a large area.
- **Squall.** Fast-moving thunderstorm or line of thunderstorms that often can produce damaging winds, hail, and tornadoes.
- **Hail.** Pieces of ice that fall from thunderstorms. Hail often is composed of concentric rings of ice that form as the particle moves through "wet" and "dry" areas of the thunderstorm.

Weather Watch. Statement about a particularly dangerous weather system that may occur at some specified time in the future.

Weather Warning. Statement that dangerous weather is likely or is occurring. Take action.

Well. A bored, drilled, or driven shaft, or a dug hole whose depth is greater than the largest surface dimension and whose purpose is to reach underground water supplies or oil, or to store or bury fluids below ground.

Wellhead Protection Area. A protected surface and subsurface zone surrounding a well or well field supplying a public water system to keep contaminants from reaching the well water.

West Antarctic Ice Sheet. See *ice sheet.*

Wetland. Land that stays flooded all or part of the year with fresh or salt water. An area that is regularly saturated by *surface water* or *groundwater* and subsequently is characterized by a prevalence of vegetation that is adapted for life in saturated-soil conditions, such as swamps, fens, marshes, bogs and estuaries.

Wilderness. An area unaffected now or in the past by human activities and without a noticeable presence of human beings.

Wildlife Refuge. An area designated for the protection of wild animals, within which hunting and fishing are either prohibited or strictly controlled.

Willy-willy. Australian term for tropical cyclone, hurricane.

Wind. A natural motion of the air; especially a noticeable current of air moving in the atmosphere parallel to the Earths' surface. Winds are caused by unequal heating and cooling of the Earth and atmosphere due to absorbed, incoming solar radiation and infrared radiation lost to space—as modified by such effects as the Coriolis force, the condensation of water vapor; the formation of clouds, the interaction of air masses and frontal systems, friction over land and water; etc.

Wind Power. Alternative energy source that has been used by people for centuries. More recently thousands of windmills have been installed to produce electric energy.

Wind Vane. An instrument used to indicate wind direction.

Window. Term used to denote a region of the electromagnetic spectrum where the atmosphere does not absorb radiation strongly.

Wood Energy. Wood and wood products used as fuel, including roundwood (*i.e.*, cordwood), limbwood, wood chips, bark, sawdust, forest residues, and charcoal.

World Ocean Circulation Experiment (WOCE). A study of the general global circulation of the oceans. It emphasizes the measurements and understanding needed to describe and understand the circulation, to simulate it, and to predict its changes in response to climatic changes.

X

Xenobiota. Any biotum displaced from its normal habitat; a chemical foreign to a biological system.

Y

Yard Waste. The part of solid waste composed of grass clippings, leaves, twigs, branches, and other garden refuse.

Yellow-Boy. Iron oxide flocculant (clumps of solids in waste or water); usually observed as orange-yellow deposits in surface streams with excess iron content.

Yield. The quantity of water (expressed as a rate of flow or total quantity per year) that can be collected for a given use from surface or groundwater sources.

Z

Zephyr. A Mediterranean term for any soft, gentle breeze.

Zero Air. Atmospheric air purified to contain less than 0.1 ppm total hydrocarbons.

Zonally-averaged Models. *Statistical-dynamical* or *energy-balance models* in which only the latitudinally averaged quantities are determined and the effects of the longitudinally varying transports are determined parametrically. Abbreviated as ZAM.

Zone of Aceration. The zone or layer above the water table in which some water may be suspended or moving in a downward migration toward the water table or laterally toward a discharge point.

Zone of Saturation. The layer beneath the surface of the land containing openings that may fill with water.

Zooplankton. That portion of the *plankton* community comprised of tiny aquatic animals eaten by fish.

